RESINOGRAPHY

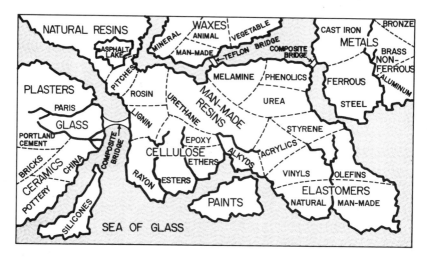

A fanciful map of the world of plastic materials. The continent of natural and man-made resins is shown associated with the three other major continents: plasters, waxes, and metals. Modeled after the map published by Ortho Plastic Novelties, Inc. in *Fortune,* Oct., 1940, and the map in *The Encyclopedia of Microscopy,* G. L. Clark, ed., Van Nostrand Reinhold Publishing Co., New York 10001, 1961, p. 526.

RESINOGRAPHY:

AN INTRODUCTION TO THE DEFINITION, IDENTIFICATION, AND RECOGNITION OF RESINS, POLYMERS, PLASTICS, AND FIBERS

By

THEODORE GEORGE ROCHOW

Associate Professor Emeritus
North Carolina State University at Raleigh

and

EUGENE GEORGE ROCHOW

Professor Emeritus
Harvard University

PLENUM PRESS • NEW YORK AND LONDON

Library of Congress Cataloging in Publication Data

Rochow, Theodore George.
    Resinography: an introduction to the definition, identification, and recognition
of resins, polymers, plastics, and fibers/ by Theodore G— R— and E— G— R—. -- New York
    xiv, 193 p.   Plenum, c1976.
    Includes bibliographical references and indexes.
    1. Resinography. I. Rochow, Eugene George, 1909-   6   joint author. II.
Title.
TP979.R6 1976   668.4'19   75-34208
    ISBN 0-306-30863-0

© 1976 Plenum Press, New York
A Division of Plenum Publishing Corporation
227 West 17th Street, New York, N.Y. 10011

United Kingdom edition published by Plenum Press, London
A Division of Plenum Publishing Company, Ltd.
Davis House (4th Floor), 8 Scrubs Lane, Harlesden, London, NW10 6SE, England

Printed in the United States of America

11/8/78 $29.50

th Rochow
stance

PREFACE

Resinography is a strange new word to many people. Like all scientific terms, it is a word coined for a specific purpose: to indicate (in this case) that resins, polymers, and plastics write their own history on the molecular and other structural levels. The word indicates further that anyone trained and equipped to ask the right questions (by means of instruments and techniques) will be able to read that history. That person must have sufficient training and experience to interpret the answers, of course, and he or she needs to have the temperament of a detective. But in the end, as readers of this book will discover, one is able to identify the material, to determine its history of treatment, and to learn much about its possible field of usefulness.

Obviously, the resinographer seeks to do the same thing with resins, polymers, and plastics that the metallographer does with metals and their alloys. Often the investigative techniques and the instruments, too, are similar, but sometimes they are decidedly different. Perhaps it would be best to say that resinography and metallography[1] (and petrography as well) share a common origin, and that origin is deeply rooted in microscopy. The "grandfather" of all three "ographies" was Henry Clifton Sorby (1826–1908),[2] who initiated metallography and petrography,[3] and was the first to report on the microstructure of a resin (amber, a natural fossil resin).[4] Indeed, Sorby also anticipated the theory of polymerization when he speculated on how a fluid exudate of a tree hardened to solid amber: "It appears as though, whilst the general mass of the resin was still somewhat plastic, a change took place — which made the material so hard that its molecular state could be permanently affected by mechanical pressure.''[4]

The purpose of this book is to organize the science of resinography so that it may be presented as a college course to graduate and qualified undergraduate students. The organization presupposes some awareness of macroscopy and microscopy, which are essential.

The plan of instruction herein is simple. First the student is introduced to the descriptive characteristics of known samples of resins, polymers, and plastics, and then (once the correlation is understood) he is encouraged to identify unknown specimens. While the emphasis is on macroscopical and microscopical description of structure and morphology, other methods of obtaining information (x-ray diffraction, electron and NMR spectroscopy, absorption spectroscopy, etc.) are treated in principle, if not in detail. The student is urged to consult the indicated authoritative references in these supplementary fields, and to use all available methods. A good detective will welcome all clues, from whatever sources.

It is indeed high time that instruction in resinography be organized and presented in book form. While America has long been eminent in metallography, and has been first to exploit plastics and synthetic fibers on a nationwide scale, our leadership in the fields of resins, polymers, and plastics may now be transferred to Japan or to Europe[5] because for a long time we had no standards for training engineers and technologists in these fields. The ASTM Committee E-23 on Resinography, for example, was not established until 1964,[6] and the development of standards has been slow. The only formal course in resinography known to the authors is the one developed in the School of Textiles, North Carolina State University at Raleigh, beginning in 1970. This book derives from the lectures given in that course over a five-year period. It is hoped that the book will inspire the teaching of resinography in other institutions as well, and will be helpful to those who have had no formal instruction in the field but find a need for its techniques in their work.

The authors acknowledge with gratitude the inspiration and instruction received from their early teachers, the late Emile Monnin Chamot and Professor Emeritus Clyde Walter Mason.[7] Many of the methods described herein were developed over a 30-year period by the senior author and his group in the laboratories of the American Cyanamid Company, and he gratefully acknowledges the Company's support as well as its kind permission to publish descriptions of methods and to reproduce some of its pictures. Credit for early help was given in the chapter which was written with R. L. Gilbert for Volume V of the famous Mattiello series in 1946.[8] Since then the following people of American Cyanamid, among others, have helped in the development of the literature on resinography: R. J. Bates,[9] M. C. Botty,[10] J. J. Clark,[11] R. E. Coulehan,[10] C. D. Felton,[10,11] J. I. Gedney,[10] D. G. Grabar,[10,12] J. P. Hession,[13,14] A. O. Mogensen,[10] F. G. Rowe,[15] E. J. Thomas,[10,11] and H. P. Wohnsiedler.[12] Grateful acknowledgment of continued administrative support at American Cyanamid is also made to Dr. G. L. Royer, and Dr. D. L. Swanson.

Another organization which contributed a great deal to the subject of this book is the American Society for Testing and Materials. It sponsored and published the first symposium on resinography,[10] and formed a national committee (E-23) on the subject.[16] We are grateful to the ASTM for permission to reproduce some of the figures and text from its publications. Personal thanks go to P. J. Smith, retired Staff Senior Assistant Technical Secretary of that organization, for his enduring interest in resinography and to S. W. Bowman, present Staff Liaison Officer to E-23 for his expert administrative advice. We thank also G. G. Cocks,[16] E. J. Pagé,[16] and R. E. Wright[10,17] of the ASTM for their outstanding cooperation.

During the past five years, while the senior author has taught resinography in the School of Textiles of the North Carolina State University at Raleigh, he has received whole-hearted support from Dean D. W. Chaney of the School of Textiles, and Dr. V. T. Stannett of the Department of Chemical Engineering. The faculty and students at N.C.S.U. have made the development of this course a rewarding enterprise, and particular thanks go to Dr. M. H. Mohamed, Dr. M. H. Theil, Dr. P. A. Tucker,[18,19] L. D. Nichols,[20] R. Heeralal,[20] L. H. Lagman, M. Kleinfelter, R. Crook, J. D. Williams III, I. A. Morrozoff and many other former students.

In a work of this kind it is impossible to include a complete bibliography, so we have had to be content with a considered choice of references. The reader may wish to consult several general publications in the field of resinography as well.[21,22] Lastly, since this is probably the first book devoted entirely to resinography, there is immense room for improvement. Suggestions to that end by the reader will be welcome.

Raleigh, North Carolina   Theodore G. Rochow
1976   Eugene G. Rochow

REFERENCES

1. *Metallography—a Practical Tool for Correlating the Structure and Property of Materials,* ASTM Special Technical Publication 557, American Society for Testing and Materials, Philadelphia, Pa. 19103 (1975).
2. D. W. Humphries, The contributions of Henry Clifton Sorby to microscopy, *The Microscope and Crystal Front* **15** 351–362 (1967).
3. D. W. Humphries, Sorby: the father of microscopical petrography, Chapter 2 in *History of Metallography* (C. S. Smith, ed.), American Institute of Mining and Metallurgical Engineers, 345 E. 47 St., New York, N.Y. 10017 (1965). See also Chapter 3, A bibliography of publications by H. C. Sorby.
4. H. C. Sorby, On the microscopical structure of amber, *Monthly Microscopical J.* **16**, 225–231 (1876).
5. Anon., U.S. lead in polymer technology threatened, *Chem. and Engineering News,* Jan. 7, 1974, 24–26, response: Dec. 23, 1974, 15; Jan. 27, 1975, 5.
6. *ASTM Yearbook 1973–74,* pp. 34, 81, American Society for Testing and Materials, Philadelphia, Pa. 19103.
7. E. M. Chamot and C. W. Mason, *Handbook of Chemical Microscopy,* John Wiley and Sons, Inc., New York, N.Y. 10016, Vol. 1, 3rd ed. (1958) and Vol. 2, 2nd ed. (1940).
8. T. G. Rochow and R. L. Gilbert, Resinography, in *Protective and Decorative Coatings* (J. J. Mattiello, ed.), Vol. 5, John Wiley and Sons, Inc., New York, N.Y. 10016 (1946).
9. T. G. Rochow and R. J. Bates, A microscopical automated microdynamometer microtension tester, *ASTM Materials Research and Standards* **12**, No. 4, 27–30 (1972).
10. T. G. Rochow, Chairman, *Symposium on Resinographic Methods,* ASTM Special Technical Publication 348, American Society for Testing and Materials, Philadelphia, Pa. 19103 (1964).
11. C. D. Felton, E. J. Thomas, and J. J. Clark, Ultraviolet microscopy of fibers and polymers, *Textile Res. J.* **32**, No. 1, 57–67 (1962).

12. *Symposium on Morphology of Polymers* (T. G. Rochow, ed.), *Interscience Div.*, John Wiley and Sons, Inc., New York, N.Y. 10016 (1963); also in *J. Polymer Sci., Part C*, **3**, iii–164 (1963).

13. J. P. Hession, D. Grabar, and F. Rauch, Thermal behavior of TNT, *The Microscope* **18**, 4th quarter (Oct. 1970).

14. J. P. Hession, G. Castillion, D. Grabar, and H. Burkhard, Polywater: methods for identifying polywater columns and evidence for ordered growth, *Science* **167**, 865–868 (1970).

15. T. G. Rochow and F. G. Rowe, Resinography of some consolidated separate resins, *Analytical Chem.* **21**, 461–466 (1949).

16. Scope and organization of committee E-23 on resinography, annual *ASTM Year Book*, American Society for Testing and Materials, Philadelphia, Pa. 19103.

17. R. E. Wright, Chairman, *Symposium on Resinography of cellular plastics*, ASTM Special Technical Publication 414, American Society for Testing and Materials, Philadelphia, Pa. 19103 (1967).

18. P. Tucker and W. George, Microfibers within fibers: a review, *Polymer Engineering and Science*, **12**, No. 5, 364–377 (1972).

19. W. George and P. Tucker, Studies of polyethylene shish-kebab structures, *Polymer Engineering and Science*, **15**, No. 6, 451–459 (1975).

20. L. D. Nichols, M. H. Mohamed, and T. G. Rochow, Some structural and physical properties of yarn made on the integrated composite spinning system. Part I: Two-component yarns, *Textile Res. J.* **42**, 338–344 (1972). M. H. Mohamed, T. G. Rochow, and R. Heeralal, Part II: Three-component yarns, *Textile Res. J.* **44**, 206–213 (1974).

21. T. G. Rochow, Resinography, in *Encyclopedia of Microscopy* (G. L. Clark, ed.), Reinhold Publishing Co., New York, N.Y. 10001 (1961).

22. T. G. Rochow, Resinography, in *Encyclopedia of Polymer Science* (H. F. Mark, ed.), Interscience Div., John Wiley and Sons, Inc., New York, N.Y. 10016 (1970), Vol. 12, pp. 66–90.

CONTENTS

DEFINITIONS, GENERAL SCOPE, AND LIMITATIONS OF RESINOGRAPHY

SOME ESSENTIALS

Science is based on *description*. Even before the scientific age, natural materials were described in terms of visible and sensible properties, such as color, odor, feel, relative density, and so on. Since most materials are complex, it soon became necessary to consider *composition* (that is, the kind and proportion of individual recognizable components present in the combination). Only then could the material be depended upon for some intended purpose, let us say building a house. These three aspects of materials are worth considering more carefully:

Description is the graphic (vivid) recording of an image or impression, either in words or by means of a picture. It may be general or detailed, and on any scale.

Composition includes the qualitative and quantitative statement of components. The source of information is analysis, and the derived facts need analytical interpretation.

Properties are characteristic qualities or quantities (numbers) which serve to identify a material in a unique way, either as a peculiarity or as an attribute common to members of a class.

In describing a room in a building we talk about the kind of construction (structure) and the shape and size (morphology). The outside walls are generally structured differently from the inside walls. In an oblong room, parallel walls have the same morphology whether or not they have the same structure. The floor and ceiling generally have the same morphology but different structure.

In talking about the composition of a house we include the condition of its materials. In talking about properties we include behavior, as (for example) the behavior of wood during seasoning, and the behavior of paint during weathering.

For reasons of greater precision we must expand the three aspects of resinography into six, as follows:

Structure is intended to include the submicroscopical arrangement of atoms, molecules, ions, or radicals within a material. That is, structure describes the *kind* of building block and the *pattern* of its arrangement, the whole being unified by intrinsic forces. A characteristic structure depends primarily on composition, and secondarily on conditions. Various intrinsic properties, such as melting point, phase-transition temperature, and solubility arise from structure.

Morphology is the arrangement of structural units into recognizable and typical shapes and sizes. Morphology depends primarily on external treatment and environment, just as structure depends on internal forces. Morphology influences most behavior, such as the *rate* of dissolving, or of melting, or of transforming into another phase, or even the rate of reacting. It is imperative that these terms (and the others in this section) be used strictly and unambiguously. If a term such as "form" can relate variously to structure, morphology, a single kind of crystal face, or a polymorphic phase, the resinographer should define what he means by the term, or else not use it.

Condition is that aspect of description which is influenced by treatment, processing, or use. It can be more important than composition at times, especially in the case of high polymers. For example consider Figure 1-1, which is from a resinographic study by Botty *et*

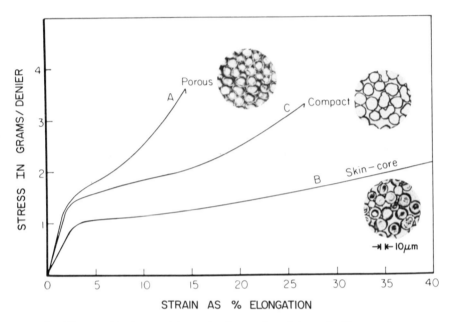

FIGURE 1-1. Resinographic study of experimental acrylic fibers, cut successively from the same wet tow and given various treatments before drying. Corresponding structures and stress–strain properties are depicted.

al.[1] of experimental acrylic fibers all from the same batch of wet tow.* The three portions, however, received very different treatments before drying. The different appearances of the fibers, and the three different stress–strain curves, represent different *conditions* of the fibers which are correlated with differences in structure brought about by the different methods of processing.

Behavior is the way characteristic properties of a substance change with time, temperature, irradiation or illumination, humidity, and other environmental factors.

---

*Tow refers here to a bundle of man-made fibers prior to spinning, as it was used earlier in connection with hemp or jute. The word comes through Middle English from the Icelandic word for wool.

The six aspects defined so far delineate the field of resinography and lead us to a definition:

Resinography is the science of structure, morphology, composition, properties, condition, and behavior of resins, polymers, plastics, and the products derived from them.

For the purposes of this book we need to define "resin," "polymer," and "plastic." The term "resin" is descriptive, the term "polymer" relates to composition, and the word "plastic" relates to properties, so a material might quite properly come under all three headings. To keep them straight, a *resin* is a solid or semisolid material which looks, feels, flows, softens, fractures, or even smells like a natural resinous substance such as shellac, rosin, or amber. A *polymer* is a substance consisting of large molecules characterized by the repetition of one or more types of monomeric units joined together by chemical bonds. Polymers may be natural, such as cellulose (cotton, linen, wood fiber), or modified, such as regenerated cellulose (rayon), or synthetic, such as polypropylene or polyurethane. A natural polymer (such as Hevea rubber) may sometimes be duplicated synthetically for economic reasons, as when polyisoprene is made from petroleum. Any other kind of rubberlike material is called an elastomer or artificial (not synthetic) rubber. A *plastic* is defined by the ASTM as a material "which at some stage in its manufacture or processing can be shaped by flow."[2] *Thermoplastics* can be softened repeatedly by heat or solvents or extramolecular plasticizers, and hence can be etched by appropriate solvents. *Thermoset* plastics harden by chemical reaction and cannot be re-softened by heat; they are relatively insoluble and can be etched only by appropriate chemical reagents which attack their structure.

THE WORLD OF PLASTIC MATERIALS

Obviously many kinds of materials are shaped by flow at some stage of manufacture or processing. A pottery vase is shaped by applying gentle pressure to the plastic clay on the potter's wheel, and afterward the dried piece is fired to a very high temperature in order to bring about the chemical reactions which make enduring earthenware out of soft clay. Many metals also are shaped by forces which direct their plastic flow, as when steel is forged, or copper is drawn into wire, or aluminum is extruded. So the world of plastic materials is far larger than the mere area of organic polymers which are shaped by pressing or blowing. It would be well at this point to consider the larger domain of plastic materials in general, and to take note of the regions where the familiar areas overlap.

The frontispiece represents a fanciful "map" on which the principal plastic materials are shown in relation to each other. There are four "continents," representing plasters, waxes, resins (including resinous and rubbery polymers), and metals.

Plasters contain solid particles which are free to move around in a liquid. After being shaped, plasters are hardened permanently by removing the liquid in some manner. Ordinary mud dries by evaporation, but the replacement of water makes mud again. If the solids are of proper composition, however, and the special mud is not only dried but fired at a certain temperature, pottery, china, or bricks are the product. Studies of why the properties of such ceramic products result from a particular composition or treatment are taken up in *ceramography*. If the solids are of Portland cement or plaster of Paris, some of the water goes into the crystallization of hydrates, which is the action that produces the hardening. Putties and paints are composed of particles in a liquid that becomes a solid resin by oxidation (e.g., linseed oil), evaporation (e.g., a lacquer), or polymerization (e.g., melamine-formaldehyde resin by baking).

Metals and their alloys* are characteristically, opaque, lustrous, conductive materials which are crystalline when solid, and which are then plastic by means of slippage in certain directions within each crystal grain. Ordinarily slippage results in work-hardening. (The

---

*Metals differ radically from other materials in that they are either elements (the simple metals, such as copper, gold, iron, aluminum, etc.), solid solutions of such elements (sterling silver, gold coins, etc.), or intermetallic compounds. The type of chemical bonding also is different: see W. L. Masterton and E. J. Slowinski, *Chemical Principles*, 3rd ed., 196–199, W. B. Saunders Co., Philadelphia, Pa. 19105 (1973).

metal or alloy may usually be softened again by annealing, that is, by sufficient heating below the melting temperature.) Cold-working is the only way that single phases of elements (Cu or Zn) or solid-solution alloys (α -brass) may be hardened, but the hardness is not very great. Somewhat greater hardness is obtained in a mixture of phases such as $\alpha + \beta$ brass. Further hardness results from precipitation of a separate phase such as iron carbide to make steel from iron.

Waxes also are essentially crystalline, and, unlike plasters and putties, they may be single phases. But waxes such as paraffin wax may also contain oil in solution. Waxiness is primarily a physical rather than a chemical characteristic. The map (frontispiece) indicates that waxes may be of mineral, vegetable, animal, or synthetic origin. Waxes sometimes are incorporated in rubbers and resins to soften them or to change the surface properties.

Resins constitute the area of greatest interest to us in this book. This "continent" includes the natural and man-made resins and polymers, and all the mixtures, copolymers, and final products derived from them. Unlike plasters, waxes, and metals, resins do not need to be crystalline in order to be plastic.

The plastic materials included in the four "continents" of the frontispiece all have the following characteristic in common: they are shaped or fabricated by applying a force which exceeds the yield value. The force may be applied by any of many processes: pulling, pushing, pressing, pounding, peening, plaiting, punching, piercing, pinching, planing, polishing, parting, peeling, forging, extruding, rolling, and swaging, to mention some. The procedure by which the force is applied leaves its telltale effect too, of course, and learning to identify these effects (and so to read more of the history of the piece) is one of the pleasures of resinography.

THE PLAN

There are at least four ways of looking at resins, polymers and their derivatives, and each involves its own way of thinking. The theoretical chemist thinks on the molecular (or macromolecular) level: he thinks of composition and the array of atoms, and of the resultant

chemical and physical properties which give rise to characteristic behavior. The physical chemist thinks in terms of *phases*, which are
physically distinct portions of matter; he has to do with structure and
morphology, with interaction of phases, and with the behavior of cells,
grains, and crystals. The colloid chemist is interested in surfaces and
interfaces, and especially in emulsions, paints, fibers, films, foils, and
foams. The engineer is interested in the bulk properties of materials,
whether they be pure substances or mixtures of many phases; he wants
to know the kind, proportion, and distribution of phases only insofar as
these affect the properties of the whole material. There may even be
phases from two or more "continents" in the material. For example,
glass fibers or flakes of metal may be embedded in a resinous polymer,
but the engineer is interested only in the behavior of the composite
material (Chapter 7).

These four attitudes, with their corresponding levels of investigation, are summarized in Table 1-1. For the resinographer each of the
four levels has merit, in that each tells us something useful about the
material, so we shall consider the levels separately in subsequent chapters. In order to do so, we must first devote a chapter to some methods
of investigation. Then, after the four chapters which correspond to the
levels of Table 1-1, we shall be in a position to consider composites in
greater detail, then some aspects of static and dynamic behavior which
correspond to visual and graphic observations, and finally some industrial applications.

TABLE 1-1. Four Levels of Study for Plastics and Polymers

Level	Of interest to	Concerned with the structure and morphology of	Involves the composition and condition of	Properties and behavior
I	Theoretical chemist	Molecule	Atoms and ions	Chemical and physical
II	Physical chemist	Phase	Internal molecules	Intraphase, grain
III	Colloid chemist	Surface	External molecules	Interphase, colloidal
IV	Engineer	Material	One or more phases	Overall, bulk

THE VISUAL PROCESS, MACROSCOPICAL
AND MICROSCOPICAL

It is said that of all we learn, a very large proportion is by seeing. It becomes important, therefore, to consider the various attributes, aspects, and concepts of visibility:

1. **attitude*** (inspiration, mood, degree of interest): "None so blind as those who will not see."
2. **experience** (memory, records): Remembering mistakes and correcting them; generalizing from isolated examples.
3. **imagination** (really a projection of 1 and 2): What am I looking for? What should it look like? What new revelation is here?
4. **resolving power:** What you are given in your eye, or pay for in a microscope or a camera; the ability to reveal closely adjacent structural details as actually separate and distinct.[3]
5. **resolution:** What you obtain as the image on your retina, and interpret with your brain.
6. **contrast:** If you cannot distinguish sufficiently between the parts, you do not see them.
7. correction for **aberrations** in the eye, camera, microscope, etc.: Employ adequate lenses, or you will not get the true picture.
8. **cleanliness:** What is *on* the lenses will confuse you about the specimen, the sample, and, perhaps, the lot of material.
9. **depth of focus:** This is the distance between near and far points which are in reasonable, simultaneous focus.
10. **focus:** Sharpen the image with your eye, camera, or microscope.
11. **illumination:** Both *kind* and *extent* are very important, and should be varied so as to gain the optimum in each case.
12. **radiation:** The radiation may be in the visible or the invisible parts of the spectrum, and may even be in the form of an electron beam.

---

*Terms printed in boldface type will be found in the Glossary, which follows Chapter 10.

13. **polarized light:** If the object is **anisotropic**, or can be made so, polarized light may be employed to increase visibility.
14. **proximity:** Get closer to see more detail. Separate parts of the image should fall on different and nonadjacent receptors of the retina, separating the details. The near limit of beneficial proximity for the unaided eye is about 25 cm.
15. **field of view:** The extent of the visible area can be an important aspect of visibility.
16. **antiglare devices:** Certain coatings on lens surfaces improve visibility of the image and reduce loss of light. Moreover, a polarizing device which has its direction of light vibration crossed with that of reflected light reduces glare.
17. **cues to depth:** Stereoscopy, shadows, perspective, and relative sizes give indications of depth.
18. **working distance:** Visibility depends not only on proximity but also on accessibility of the object. Lenses of high resolution often have limited working distance.
19. **depth of field:** Roughness of surface or thickness of layer may limit the visibility.
20. **structure** of object: Sometimes the kind of construction causes diffraction patterns or interference colors, which can be either informative or misleading, depending on experience.
21. **morphology** of object: Visibility also depends upon the size, shape, and spacing of particles or parts.
22. **information** about sample: Try for all the information you can get; retrieve it from all media available. You never know which clue will turn out to be the most important!
23. **experimentation:** Test your interpretations of what you see. Confirm or refute your conclusions yourself, before others do.
24. **preparation of the specimen:** Do as little as necessary to the sample, but do it.
25. **behavior:** Look for changes in your specimen and sample, due to age, preparation, etc.
26. **photography:** This technique has its own advantages (e.g., contrast) over visual observation, and also its limitations, especially in resolution.

SOME HINTS

Sampling and Preparation. If possible, do your own sampling and get your information at the source. It may be that you can examine the material nondestructively right there, and even be able to solve the problem on the spot. If you must take your sample from someone else's hands, first examine the specimen as received, without any preparation. Remember that anything you do to a specimen will change it, possibly for the worse. Work on only a portion of the specimen at a time; if you spoil it, there will still be something left to go on. After examining a specimen as it arrived, clean it if necessary, and save the dirt. It may be important. If you break, saw, drill, or shave a specimen, save the fragments until you are sure that you will not want them. Label or engrave the specimen with a coded designation and date. Organize a system of filing samples *before* you start collecting them.

Illumination. We all know how difficult it is to see details of dark things at dusk, or of bright things in glaring sunlight. Yet, in macroscopy as well as microscopy, many do not take the trouble to get optimum illumination. Control the intensity of the light, and orient it to get the optimum angle and position of incidence. A scratch, for example, is most visible in a single narrow beam of light directed at a grazing angle and coming in perpendicular to the direction of the scratch.

Be Curious. Chamot's rule is still good: "Take a quick look. If you do not see anything, you have not wasted much time." If you do not look, you may miss everything.

Photography is the most graphic and vivid method of recording. Photographs and photomicrographs taken under the resinographer's *selected* and recorded conditions of preparation, illumination, resolution, etc. also may provide information over and above that seen visually. Furthermore, in the case of very fast or very slow action, cinematographs or time-lapse pictures provide the only way to see what is going on. Even when the emission from the sample is invisible, as with radioactive samples, or when x-rays are involved, photography provides a more satisfactory visual medium than a luminescent screen because in this case the photographic materials have a higher limit of resolution. Generally, but not always, photographs are preferred over tracings, drawings, and models as evidence. However, all methods of depiction depend on the skill and training of the observer, and all carry the integrity of the investigator with them.

Resolution of detail in the finished photograph depends not only upon the visual acuity of the observer, but also upon the inherent factors which characterize the photosensitive materials (and even the development of them). In photographic (rather than visual) examination of a specimen under equal optical conditions, the surfaces need to be smoother and the sections need to be thinner, because there is less depth of focus in the camera than in the eye. Furthermore, scratches and knife marks are much more noticeable (and permanent!) in a photograph than in visual observation, so again the preparation needs to be refined for photography.

Lastly, photography is generally an important medium for correlating *changes* in properties or behavior with changes in composition or treatment. Figure 1-2 provides an example: the mechanical load on

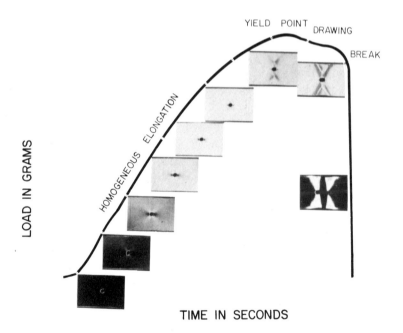

FIGURE 1-2. Load-time curve plotted automatically on the microdynamometer (Chapter 9) with pictures by cinematography corresponding to the times shown by the selected points on the graph.

a plastic foil is plotted as ordinate against time as abscissa, and each point on the curve is illustrated by a photomicrograph taken at that time. Thus the changes in structure of the foil, as manifested by the polarization pattern (originally photographed in color), are correlated precisely with the coordinate changes. Both the curve and the photographs were produced by the microdynamometer[4] which is described in Chapter 9.

SUMMARY

Resinography is a precise science, and it is necessary at the outset to consider how we may collect precise and dependable information, how we may organize and store it, how we may interpret our observations, and how to present our conclusions. In the course of doing these things, some words like *morphology, structure, composition, condition*, and, of course, *polymer, plastic*, and *resin* have to be used with exact meanings, so these terms have been defined and explained. A much more complete listing of all the key words used in this book, together with their accepted definitions, will be found in the Glossary which follows Chapter 10.

The adjective *plastic* applies to any substance that can be shaped by the application of suitably directed force, and so plastic materials include resins, polymers, elastomers, plasters, waxes, and even metals. Some relations between all these organic and inorganic materials are indicated in the whimsical "map" which appears as the frontispiece. The chief characteristics of each major class have been explained in this chapter. Plaster, wet clay, mortar, and cement mixtures are plastic by virtue of their liquid content, and are more fluid when they contain more liquid. The solid particles hinder the free motion of the liquid under small stress, and so increase the overall viscosity of the mass. When the liquid is removed by evaporation, the solid residue may harden by reaction with gases of the air (as in lime mortar), or by hydration (as in Portland cement), or by interaction of its components at much higher temperatures (as in the firing of ceramic products).

Metals are distinctive, often elementary, materials which rarely are liquid at room temperature (Hg, Cs, Ga, and alloys of Na and K) and which are crystalline when solid. The crystals may be deformed by suitable force, whereupon entire layers of atoms slide over other layers; the disordered region then resists further deformation and the metal is said to be work-hardened. The deliberate introduction of discontinuities in the form of precipitated particulate matter, such as iron carbide in steel or copper silicide in aluminum, acts similarly to harden and strengthen the metal.

Waxes are **unimeric**, crystalline, organic substances of animal, vegetable, or mineral origin. They may be natural or synthetic. Their soft crystals deform easily, and may be smeared out into a thin, slippery layer by buffing. Some waxes are compatible with organic polymers and elastomers, and may be added to them to soften them or to provide a waxy surface.

Natural resins of recent origin, such as rosin and shellac, are unimeric organic compounds. Fossil resins, such as amber, and man-made resins, such as phenol-formaldehyde, are polymeric organic compounds (organometallic in the case of the silicones) which may also be natural or man-made, and may be of animal, vegetable, or mineral origin. If man-made, they may be produced by the polymerization of a single **monomer** (to make a homopolymer) or by interpolymerization of two or more monomers (to make a copolymer). The most desirable physical properties are strength, toughness, durability, and ease of fabrication by pressing, extruding, rolling, drawing, or blowing the hot material. Polymers that harden to an infusible, insoluble mass in a heated mold are called thermosetting, and those that continue to soften with heat are called thermoplastic. Polymers are sometimes used alone as plastic materials, but more often they are combined with fibers or powders, or with other reinforcing agents, to increase their strength and durability. Pigments, dyes, antioxidants, plasticizers, or other modifiers may also be incorporated. In common parlance the noun "plastic" refers to such a modified or composite material, embracing all of the additives and reinforcing agents, rather than to a pure polymer. Although glass is thermoplastic and porcelain is thermoset, these are not commonly called plastics because they are inorganic rather than organic in their composition.

Resinography considers plastics and resins from four fundamental points of view: (1) at the level of atoms and molecules, considering the chemical bonding and resultant molecular structure; (2) at the level of crystals, grains, and phases, considering the structure and morphology determined by these units; (3) in terms of the surfaces and interfaces between phases, as these affect the colloidal behavior and the macrostructure; and (4) in terms of the overall engineering properties of the material, where measurements of bulk properties take precedence over composition, microstructure, and everything else. Each point of view involves its own techniques of investigation, which produce equally valuable information; the wise resinographer will welcome facts from all such valid sources.

Besides information from these four levels of study, and even in the absence of all other information, the resinographer must make his own examination and arrive at his own conclusions about the identity and utility of a plastic or fiber. This first chapter concludes with a consideration of 26 aspects of visual examination, listed in the approximate order of their importance. The list warrants a second thorough reading at this point. Some of the terms which appear there may be new or unclear; the Glossary will help, and a book on elementary microscopy may be necessary. The chapter concludes with some general hints about sampling a material and then lighting, examining, and photographing the specimens.

SUGGESTIONS FOR FURTHER READING

D. R. Uhlmann and A. G. Kolbeck, The microstructure of polymeric materials, *Scientific American*, Dec., 1975, pp. 96–106.

Materials, a paperback reprint ($2.50) of the Sept. 1967 issue of *Scientific American*, W. H. Freeman and Co., San Francisco, Calif. 94104 (1967). Thirteen articles on: materials, solid state, metals, ceramics, glasses, polymers, composites; thermal, electrical, chemical, magnetic, and optical properties; competition of materials.

McGraw-Hill Encyclopedia of Science and Technology, 3rd ed.,

McGraw-Hill Book Co., New York, N.Y., 10020 (1971). This is a good general reference to consult on any operation or material covered in this chapter, and to fill deficiencies in background.

Article on resinography and especially its cross references to related articles, in *Encyclopedia of Polymer Science and Technology* (H. F. Mark, ed.), Vol. 12, John Wiley and Sons, Inc., New York, N.Y. 10016 (1970).

REFERENCES

1. M. C. Botty, C. D. Felton, and R. E. Anderson, Application of microscopical techniques to the evaluation of experimental fibers, *Textile Research J.* **30**, 959–965 (1960).
2. *ASTM Glossary*, American Society for Testing and Materials, Philadelphia, Pa. 19103 (1973).
3. E. M. Chamot and C. W. Mason, *Handbook of Chemical Microscopy*, 3rd ed., John Wiley and Sons, Inc., New York, N.Y. 10016 (1958); H. Freund, *Handbuch der Mikroskopie in der Technik*, Vol. 6, Parts 1 and 2, Umschau Verlag, Frankfurt a. M., Germany (1972).
4. T. G. Rochow and R. J. Bates, A microscopical automated microdynamometer and microtension tester, *ASTM Materials Research and Standards* **12**, No. 4, 27–30 (1972).

METHODS OF INVESTIGATION

Since resinography is literally the graphic description of resins, polymers, and their products, any scheme of investigation which produces a graphic record is important here. The record could be simply a handwritten one, or typed on a punch card as in Figure 2-1. For easier retrieval, the information could be punched on tape to be fed to a computer memory. Rather than a written record, a pictorial one is often more exact and more compact, since it is the traditional equivalent of a thousand words (see Figure 2-2, which is a photograph showing the structure and morphology of a wartime automobile tire). Finer detail can be recorded by resorting to microscopy, as is evident in Figure 2-3, a montage of three photomicrographs which show the fine structure (at appropriate resolution) of an entire cross section of a commercial artificial leather. Sometimes a cinematograph (a succession of frames on a moving-picture film) provides the record, as in Figure 1-2 of the preceding chapter. And of course, the photomicrograph may be obtained at higher resolution by using an electron microscope, as in Figure 2-4, which shows the very fine structure of an acrylic fiber.

The high quality, convenience, and ready availability of modern photographic materials have made hand drawing almost obsolete, but the student would do well to develop a habit of sketching what he sees, especially during the ''quick look'' recommended in the last chapter. It not only is quicker and more convenient than setting up for photog-

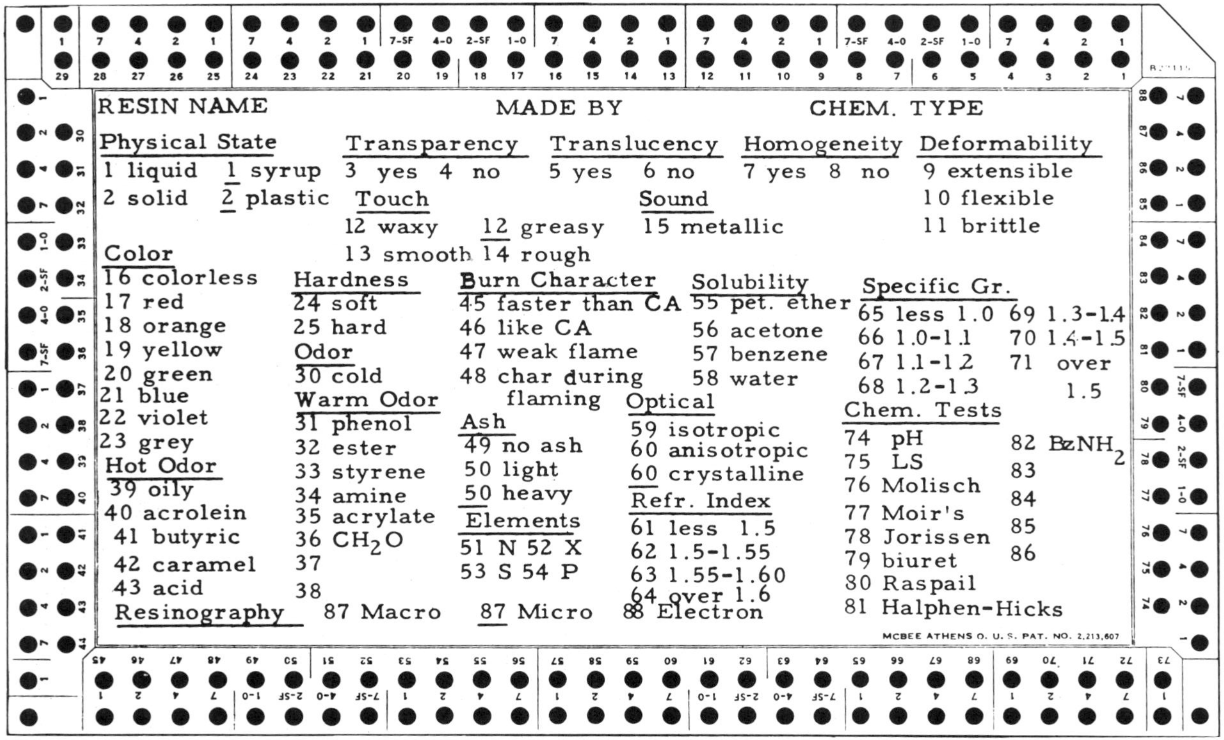

FIGURE 2-1.   Format of punch card for resinography, designed by A.F. Kirkpatrick.[1] Card is 5 by 7 in. in size.

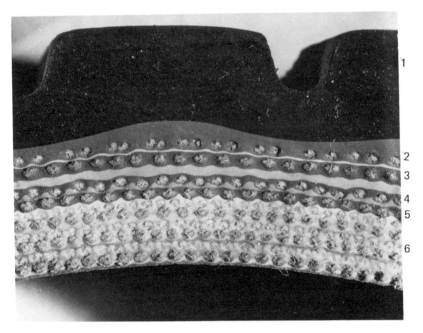

FIGURE 2-2. Cross section of a wartime automobile tire, showing construction and indicating parts of sample for identification of kinds and proportions of rubbers and fibers. (1) Tread. (2) Breaker stock. (3) Breaker cord. (4) Cushion stock. (5) Cushion cord. (6) Carcass cord and stock.

raphy, but often can emphasize particular features by simplifying the hard-to-describe details. As an extreme example, Figure 2-5 is an interpretive drawing of the appearance of the fracture surface of a thermosetting resin as seen under the electron microscope at very high resolution.

Sometimes observations of an entirely different sort give rise to a mental picture, and it is worthwhile to record that *imaginary* picture at the time. Figure 2-6 shows one polymer chemist's picturization of his temporary idea concerning the aggregation of polymer molecules in a crystalline plastic, and Figures 2-7 and 2-8 show still more abstract representations of polymer structure.

Mechanical pen tracings are a common method of recording information, not only in spectrophotometers and other spectrographic

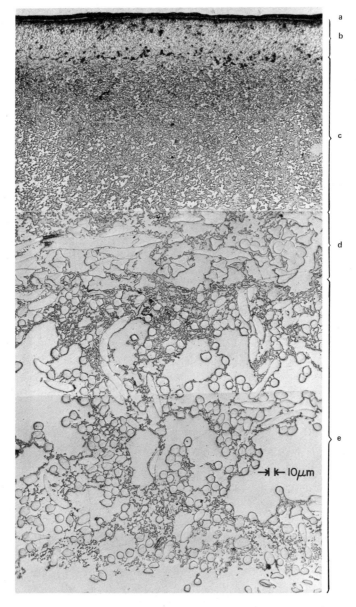

a
b
c
d
e

→ k— 10μm

FIGURE 2-3. Polished, impregnated cross section of artificial leather, by
E. J. Thomas, American Cyanamid Co. (a) First layer. (b) Pigmented resin.
(c) Finely porous zone. (d) Intermediate zone, fibers mostly trilobal in cross
section. (e) Open, porous structure.

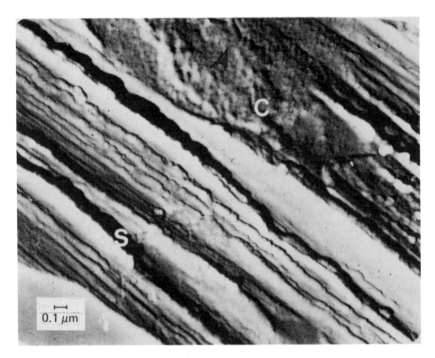

FIGURE 2-4. Electron micrograph of a replica of an experimental acrylic fiber showing the orientation of the particles in the skin (S) and randomness in the core (C). Here the resolution is sufficient to depict the macromolecules.[2]

FIGURE 2-5. Hand-drawn topography of fracture surface of M–F thermoset resin, interpreted from electron micrographs by H. C. Wohnsiedler,[3] American Cyanamid Co.

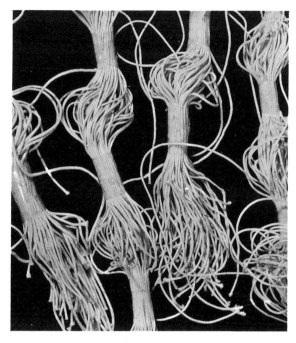

FIGURE 2-6. Imaginary, oversimplified picturization of aggregation of polymer molecules in a crystalline plastic. Each piece of string represents a molecule; the bundles represent elementary fibrils. By W. O. Statton.[3]

FIGURE 2-7. Possible defect structures in polymer lattices. The zigzag represents the CH_2 groups in a simple polymer chain. By W. O. Statton.[3]

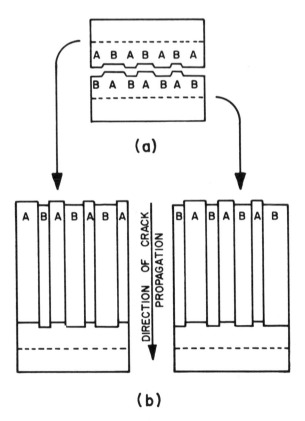

(a)

(b)

FIGURE 2-8. Model of the fracture surfaces of poly(methyl methacrylate). In the (a) and (b) regions the surface layer is of different thicknesses and hence displays different interference colors. By J. P. Berry.[3]

equipment, but also in other devices. Figure 2-9 shows the tracings recorded by a sensitive profilometer as its sensor travels over the surfaces of some well-known fibers. Figures 2-10, 2-11, and 2-12 show spectral absorption curves recorded for samples in the visible, ultraviolet, and infrared regions. The various curves of Figure 2-10 analyze the color of a polyester plastic after outdoor exposure at different times of the year; the record of Figure 2-12 is intensified by blacking in the absorption bands for emphasis. The results of gas or vapor chromatography are also recorded as a tracing (Figure 2-13), as

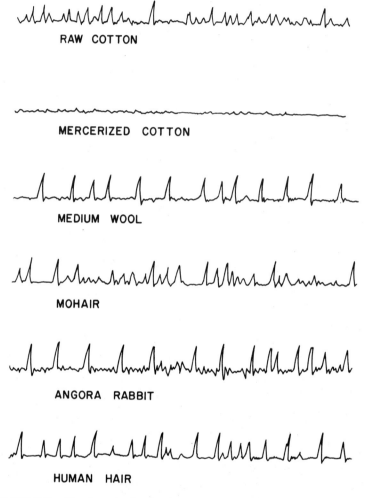

RAW COTTON

MERCERIZED COTTON

MEDIUM WOOL

MOHAIR

ANGORA RABBIT

HUMAN HAIR

FIGURE 2-9.   Tracing obtained with fiber-surface analyzer. By S. C. Schier and W. J. Lyons.[4]

are those of differential thermal analysis (Figure 2-14). In each case important information about the behavior of a polymer or plastic is written down for reference and comparison.

Mechanical tracings represent only the end result of some complicated processes that go on within the instruments that are used to analyze polymeric materials, of course, and the purposes of this chapter include a short consideration of each of the common methods used for the investigation of such materials. The discussion can be organized by considering the methods which may be employed at each of the four levels described in Chapter 1 (see Table 1-1): the study of molecules, the study of phases, the study of surfaces, and the study of overall behavior of bulk samples.

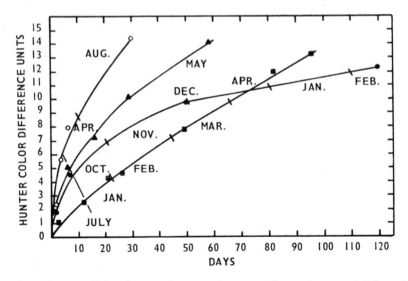

FIGURE 2-10.   Color change of a polyester vs. outdoor exposure at 45° south with months of exposure noted on curve, by Paul Giesecke, American Cyanamid Co.

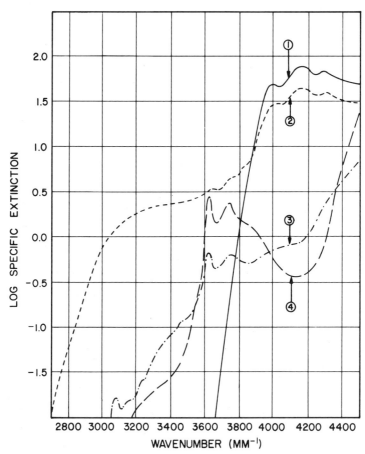

FIGURE 2-11. Ultraviolet absorption spectra of rosin and modified rosins: (1) Abietic acid. (2) WG wood rosin. (3) Hydrogenated methyl abietate. (4) Dehydroabietic acid. By the late Robert Hirt, American Cyanamid Co.

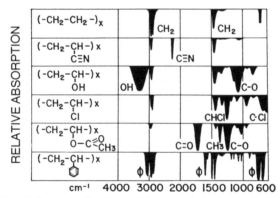

FIGURE 2-12. Schematic drawing of characteristic absorption in the infrared by some typical polymers. Courtesy of N. B. Colthup, Infrared Group, Central Research Div., American Cyanamid Co.

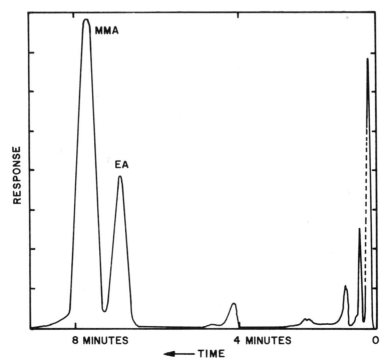

FIGURE 2-13. Chromatogram of pyrolysis products of a plastic film showing proportion of ethyl acrylate (EA) and methyl methacrylate (MMA). By R.A. Landowne, American Cyanamid Co.

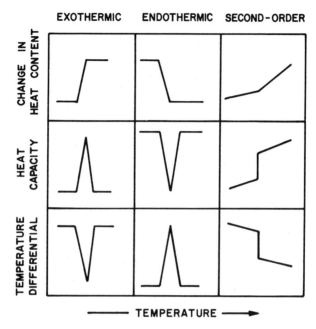

FIGURE 2-14.   Typical graphs obtained by differential thermal
analysis. By Ke.[5]

LEVEL I. THE MOLECULAR LEVEL

The best-known method for detecting the component groups and
radicals within an organic polymer or resin is *infrared spectroscopy*
(IR).[6] Here, an intense ray of infrared energy is generated by an in-
candescent body and dispersed into its component wavelengths by a
monochromator based on a prism (usually of monocrystalline NaCl,
"rock salt") or a grating, acting in conjunction with a suitable collimator which renders the rays parallel. The resulting narrow band of
infrared radiation is passed alternately through a cell containing the
sample, and then through an empty cell, as the monochromator is
operated mechanically over its range of wavelengths. The emergent
infrared beam is then picked up by detectors (which operate on

photoconductive or photovoltaic principles), and the two output signals are compared by the circuitry so that only the differential absorption of the sample is recorded as an output current. The slits, the dispersing element, and the recorder are all operated by the same mechanism, so that a plot of relative absorption vs. wavelength or frequency is obtained (see Figure 2-12).

Chemical compounds (including polymers and plastics) absorb infrared energy as a result of internal friction. The component atoms have characteristic atomic masses, and these atoms are attached to each other by chemical bonds which have characteristic strengths (called bond force constants). The resulting structure can be looked upon as an assembly of weights connected by springs. Once set in motion, the component parts will oscillate about mean positions at a frequency which is determined only by the masses of the weights and the stiffness or strength of the springs. The oscillations may be of a simple back-and-forth type (translational), or torsional (rotational), or rocking (bending), or any combination of these. In the compound, once the characteristic frequency of each type of motion is settled by the atomic masses and the bond force constants, input energy of that particular frequency will set the atoms in resonant vibration. Since the sample of polymer or plastic is not an ideal gas, but rather a very real solid (or solution of that solid), its atoms are attracted to neighboring atoms. Indeed, these same intermolecular forces are responsible for the polymer's mechanical strength. Hence the vibrating atoms are somewhat constrained by their neighbors, and their motions are subject to a kind of internal friction which gradually converts the vibrational energy to heat. Hence we have an assembly of atoms which absorbs infrared energy at characteristic frequencies, only to lose it as heat and absorb more. The result is a characteristic pattern of continuously absorbed infrared frequencies, by which we may recognize particular bonds and groupings within the molecular structure.

The recognition of polymers by their IR absorption spectra depends upon previous examination of polymers of known composition and structure, just as people may be identified by their fingerprints once those fingerprints have been put on file. It is not necessary that each resinographer compile the file of known spectra himself, for there are published compilations available.[7,8,9] After some experience in reading spectra of pure compounds, the investigator will be able to recognize the absorption bonds which correspond to common substituent groups, such as phenyl, carbonyl, siloxane, etc. The next step then is to estimate the relative quantitative proportions of the constituents from relative intensities of absorption. Details can be found in books on instrumental analysis.[10]

Nuclear magnetic resonance (NMR) provides a method for studying structure and bonding in a material by means of the *magnetic* behavior of pertinent nuclei in its constituent atoms. Protons and neutrons have magnetic moments, just as electrons do, although the moments of nucleons are much smaller. If a particular nucleus contains an even number of neutrons *and* an even number of protons, as in $^{12}_{6}C$, $^{16}_{8}O$, or $^{32}_{16}S$, there is zero net magnetic moment because the nucleons are paired, but if there is an *odd* number of protons or neutrons, there will be a resultant nuclear magnetic moment. The nucleus of common hydrogen, $^{1}_{1}H$, has one of the highest such moments, and so lends itself readily to NMR investigation. If we think of isolated hydrogen atoms (or any other atoms with nuclei having magnetic moments) as tiny bar magnets or as spinning spheres, we see that in the absence of any external magnetic field, the tiny magnets may adopt any orientation they please, but if an external magnetic field is imposed, the magnets will interact with it. Since protons have a spin quantum number of ½, the protons may align themselves either with the external field or against it; only these two states (parallel and antiparallel) are possible. The energy required to flip over a proton (and so raise it to the upper, or antiparallel, state) is proportional to the nuclear magnetic moment and the strength of the external magnetic field. At a field strength of 10,000 gauss, the frequency corresponding to the transition energy lies in the shortwave radio range. Therefore, if we place a sample containing protons in a magnetic field of several thousand gauss and irradiate it with radio-frequency energy of gradually increasing frequency, at the exact resonance point there will be an absorption of energy. The effect is very small because the nuclear moments are small, but by suitable pulsing techniques the absorption may be made very clear.

In a rigid crystal lattice the neighboring protons in fixed positions exert their own magnetic fields, which are added to or subtracted from the external field, and so the effect is smeared out. In a liquid, however, the nearest neighbors change places rapidly and the local fields cancel each other. Hence we might expect a single sharp absorption line for such a hydrogen-containing liquid. The electron density provided by covalent bonds exerts a very minor shielding effect, however, and so there are minute chemical shifts of the resonant frequency depending on the local electron density of the H—C or H—O bonds in the particular organic compound being studied, shifts in the order of several parts per million.

The NMR spectrum of an organic compound shows a pattern of proton absorption lines which may be characterized by referring to the spectra of known compounds.[11] It is necessary only that the sample be liquid or be dissolved in a liquid noninterfering solvent (such as CCl_4). Instruments of high refinement are expensive, but they produce NMR spectra of great detail and reliability from which the molecular structure may be deduced (see Figure 2-15, which compares the NMR spectra of **isotactic** and **syndiotactic** poly(methyl methacrylate). Polyesters give very distinct spectra from which the types and proportions of starting materials can be deduced. As in infrared spectroscopy, many spectra of pure compounds have been published, and there are books on the interpretation of the spectra of unknown samples.[11-16]

Electron spin resonance (ESR) detects radicals trapped in rigid polymers. The radicals may be some left from the polymerization-initiating process at the time the polymer gelled, or they may have been formed by irradiation (Figure 2-16). The principle involved is similar to that employed in NMR, except that the magnetic moment of an unpaired electron is much greater than any nuclear moment.[17] Electron spin resonance can provide information about the disruption of structure caused by sunlight or by radiation, and can also measure the effectiveness of antioxidants and stabilizers which are supposed to quench the radicals produced by such irradiation, under various conditions of temperature and concentration of oxygen. Furthermore, since the unpaired electron of a free radical provides a strong localized magnetic field of its own, such an electron will interact with the neighboring magnetic nuclei to produce a characteristic spectrum (see Figure 2-16).

Ultraviolet (UV) *absorption spectroscopy* involves changes in the electronic energy levels of molecules from the ground state to an

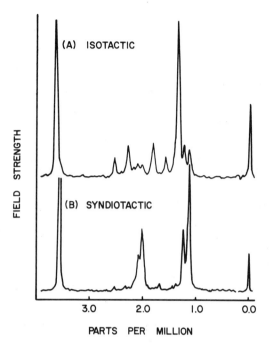

FIGURE 2-15. Nuclear magnetic resonance (NMR) of isotactic vs. syndiotactic poly(methyl methacrylate) (PMMA), by Lancaster *et al.*, American Cyanamid Co.

excited state, and the consequent relaxation. An intense source of ultraviolet light (usually a mercury-vapor or hydrogen discharge lamp) provides a collimated beam which is dispersed by a grating, while a constantly changing band of UV frequencies is passed through the sample contained in a fused-silica cell. The transmitted ultraviolet radiation is measured by a photoelectric detector, and the output is shown by a recorder coupled to the monochromator mechanism. As in IR spectroscopy, provision is made for plotting only the differential absorption (subtracting the absorption of an empty cell and compensating for the nonlinear output of the source and the monochromator). The characteristic absorption band of a particular grouping of atoms (known as a chromophore, as in dye chemistry) serves to identify that group and to follow its behavior as various antioxidants or UV absorbers are added. Such spectrophotometry in the ultraviolet region is a valuable nondestructive technique for detecting and estimating double or triple bands in organic substances. Figure 2-17 shows the absorption spectra for standards of double, triple, and quadruple conjugation in such unsaturated compounds, and Figure 2-18 shows differences in conjugation for various drying oils used in paints and varnishes. Figure 2-11 shows the differences in UV absorption of normal rosin and modified rosin. Ultraviolet spectrophotometers are relatively simple and inexpensive, and they can provide much useful information. Again, there are compilations of spectra of known compounds.[18] Further details can be found in practical manuals.[19-21]

Mass spectrometry is an entirely different technique of investigation in which substances in the vapor state (such as products of thermal

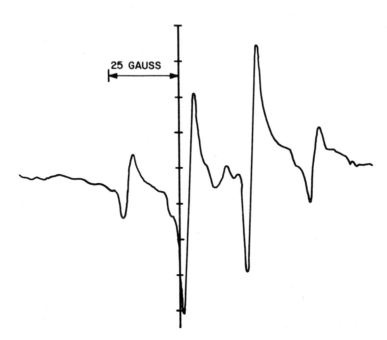

FIGURE 2-16. Electron spin resonance of PMMA in liquid N_2, UV irradiated. Unpaired electron is interacting with magnetic nuclei (H of CH_3), by Lancaster *et al.*, American Cyanamid Co.

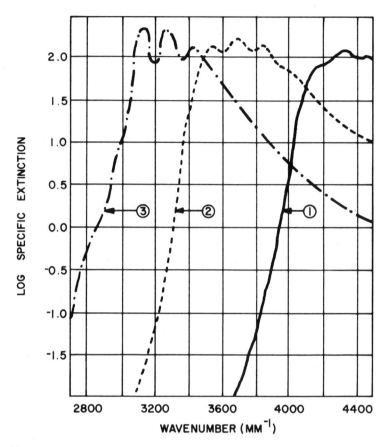

FIGURE 2-17.   Standards for ultraviolet conjugation assays: (1) Double conjugation, 10,12-linoleic acid, $CH_3 (CH_2)_4(CH=CH)_2-(CH_2)_8 COOH$. (2) Triple conjugation, eleostearic acid, $CH_3 (CH_2)_3(CH=CH)_3-(CH_2)_7 COOH$. (3) Quadruple conjugation, parinaric acid, $CH_3CH_2(CH=CH)_4-(CH_2)_7COOH$. By the late Robert Hirt, American Cyanamid Co.

decomposition or of degradation) are bombarded by a stream of electrons of known energy, causing bond rupture and ionization of the fragments. The positively charged ions then enter a region of applied magnetic and electric fields (within an evacuated tube), and these fields are then varied progressively so that ions of successively greater mass traverse the evacuated space and strike a collector, which picks up and amplifies their electric charges. Just as any moving object of light weight (a feather or a piece of paper) is deflected from its path more easily than a heavy object (a speeding car or a running football player), so the lighter ions are deflected at lower magnetic and electric fields than are the heavier ions. By recording the ion current (the *quantity* of charge) at each successive increment of field strength, a quantitative record of fragments (a "mass spectrum") is obtained.

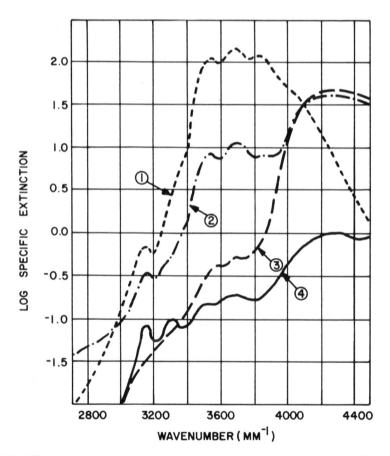

FIGURE 2-18. Ultraviolet absorption curves for some typical oils: (1) Tung oil (largely triple conjugation). (2) Isomerized linseed oil (mixed triple and double conjugation). (3) Dehydrated castor oil (double conjugation). (4) Linseed oil (little conjugation). By the late Robert Hirt, American Cyanamid Co.

It is to be expected that a given material will break down into ionized fragments in the same way whenever a sample of it is bombarded with electrons of the same energy under the same conditions. Hence the pattern of decomposition fragments (its mass spectrum) will be characteristic of that material, and we can recognize the material even when it is mixed with other substances. The resinographer can build up a file of mass spectra of known compounds and pure polymers, and after some experience with known mixtures, he can learn to identify the components of an unknown. The chief difficulty arises from the necessity of using a vaporized sample. A solid material can only be converted to vapor by pyrolysis or depolymerization, and the conditions of this conversion become the critical factors. Details are given in various handbooks.[22-24]

Molecular weights can be determined by means of a mass spectrometer, of course, if the substance is volatile. Often, however, we are dealing with a nonvolatile polymeric substance or with a mixture of substances. The usual methods for determining the molecular weight of a pure compound (vapor density, elevation of the boiling point, or depression of the freezing point of a solvent) are not suitable for a collection of large molecules or a collection of molecules of widely different sizes and weights. The molecules of a polymer vary so widely in morphology that special methods must be used to determine their weight.

The problem is even greater than this, because the molecules of different polymers have different *ranges* of weight, different *distributions* of molecular sizes and weights, and different *average* sizes or weights. The easiest figure to obtain is the average molecular weight. But even here there are several kinds of averaging we can do: we can get a number average $\overline{M}_n$, or a weight average $\overline{M}_w$, or a viscosity average $\overline{M}_v$. The number average comes out of any technique based on colligative properties, such as determination of osmotic pressure, because every molecule contributes equally to the osmosis, regardless of size; the result depends only on the number of molecules in the sample. A weight average, on the other hand, comes out of a measurement of light scattering, because each molecule makes its contribution to the scattering effect *in proportion to its size or weight*. The weight average is a larger figure than the number average because the larger molecules in the mixture make a greater contribution. The ratio of the weight average to number average is taken as a measure of the breadth of the molecular weight distribution.

Where does the term "viscosity average" come in, then? The viscosity of a solution depends in still a different way on the amount and kind of solute present, and so if we determine the *intrinsic* viscosity of a polymer solution, we get from it an average molecular weight which is not the same as $\overline{M}_n$ or $\overline{M}_w$; it is somewhat less than $\overline{M}_w$ but is nearer to it than to $\overline{M}_n$.

It is not necessary that the very important *distribution* of molecular weights be estimated only from the ratio $\overline{M}_w/\overline{M}_n$; it can be determined directly by the technique of gel permeation chromatography (GPC).[25] To a first approximation, the weight of a particular molecule is exponentially related to the volume of solvent required to elute it from the column material that holds the gel. The resulting chromatogram is a distorted curve of molecular weight distribution. In principle, the distortion can be corrected by adequate calibration, but in practice the correction is difficult. Nevertheless, the GPC method is a rapid and automatic one for comparing the breadth of distributions of molecular weights in samples of polymer, if they are of identical chemical composition.

Equilibrium centrifugation is still another technique, and one which offers the greatest hope of obtaining the molecular weight distribution from a single experiment with *soluble* polymers. The theoretical foundation has been established, but the experimental technique and the calculation may need refinement. Briefly, the idea is to spin the solution in an ultracentrifuge at such a speed that the centrifugation pattern remains static (that is, the molecules are not removed from the solvent but are held at a point where the centrifugal force tending to remove them is balanced by the dispersive forces due to thermal agitation).

It should be emphasized that *all* of the methods for determining molecular weight listed so far depend upon a solution of the sample in a suitable solvent. Such methods cannot be applied to an insoluble material like a thermoset polymer, nor can they be used to investigate tiny or microscopic samples. For *in situ* examination of such small samples (or of insoluble ones), electron microscopy offers some hope, but only if the molecules are giant ones with weights in the hundreds of thousands or millions.[4] In these cases the distribution is obtained by direct observation and measurement of the giant molecules or their replicas on the electron micrograph at known magnification. Here we are counting and measuring intangibly separate molecules in a selected phase, within a very small region. The principal limitation, other than the laborious and painstaking technique, is the very tiny sample which must be taken as representative of the whole. On the other hand, the separate particles of a particulate phase can be examined one at a time to look for differences.

LEVEL II. THE LEVEL OF PHASES. THE DIFFERENT KINDS OF MICROSCOPES

A phase is a homogeneous, physically distinct and separable portion of matter. The "separable" aspect means only that it must be tangibly or intangibly distinct, so that it could be touched or pulled out of a mixture if we only had tools small enough to do that.

The various microscopes, including the electron microscope, are instruments for determining the kinds, proportions, and distribution of phases on a surface or in a material. Microscopical examination can be either qualitative or quantitative; formerly it was largely qualitative, but recently more and more quantitative microscopical analysis is accomplished by automatic scanning of the sample. This means that an electronic "eye" (receptor of some kind) must be able to distinguish one phase from all of the rest. It calls for *intangible separation* on the basis of grayness, color, texture, or other distinctive property. By a simpler method, a linear interception of one kind of boundary can be measured by means of cross-hairs in the eyepiece and a micrometer screw and scale on the stage. In a special recording micrometer integrating stage, there is a separate recording for each of 4, 6, or more constituents. The selection of categories formerly was done by human eye and brain, but nowadays some scanning, selection, and computation are electronic and automatic.

The recognition, identification, and description of phases may also be performed intangibly by means of the polarizing microscope. By relying on the different ways in which the different phases affect a beam of polarized light, it is possible to distinguish the phases and to determine several optical properties of each. It is possible also to watch a particular phase while the sample is subjected to various influences, and so determine the melting point, phase-transition temperature,[26] hardness, rate of crystallization, and critical solution temperature.[27] As for the tangible separation of phases, that can be accomplished by gel permeation chromatography,[25] adsorption chromatography,[28] thin-layer chromatography, liquid–liquid chromatography,[29] differential solubility, fractional crystallization, or any other standard method for separating and purifying phases.

The *crystalline structure* of a phase may be determined macroscopically by x-ray diffraction, usually by obtaining and interpreting a "powder pattern" from a sample of unoriented polycrystalline powder. The pattern is compared with those of known polymer samples kept on file, or with those collected and made available by the ASTM. If a full determination of crystal structure is to be made, in terms of the crystal system, the space group, and the dimensions of the unit cell, it is necessary to use a single crystal a few tenths of a millimeter or more in diameter.[30]

If only very small crystals are available, or if only small samples of fine powder are available, these may be subjected to the x-rays generated in a transmission electron microscope. The patterns are the same as those obtained macroscopically, so the same reference sources are applicable. An advantage of the electron microscopical method is that an interesting crystalline phase among a group of phases may be singled out intangibly and investigated without removing it.

LEVEL III. METHODS FOR EXAMINING SURFACES

Sometimes the surface of a resinographic specimen can be examined just as received, but more often the surface will need polishing, etching, or some other preparation. Reflected light is used, with or without a microscope, depending on the resolving power required to reveal the interesting and important details (Figure 2-19). Similarly, reflected infrared radiation may be used, if an IR spectrograph with reflection optics is available, or reflected ultraviolet light with a UV microscope. Even reflected electrons or x-rays may be used, at times, in which case a diffraction pattern or an interference graph is obtained (see Figure 2-20). It is only by using *reflected* radiation (of whatever kind) that one can be sure that the *surface* produces the effects and not the interior of the material.

Similarly, to examine the surface itself at highest resolution one must use *reflected* electrons, not transmitted ones. The instrument involved here is the scanning electron microscope (SEM), and it is quite different in principle from the more usual transmission electron microscope. The scanner consists of an electron source and a magnetic lens which projects a greatly reduced image of the source on the specimen. This image is moved systematically over the surface of the specimen by vertical and horizontal sweep circuits, somewhat in the

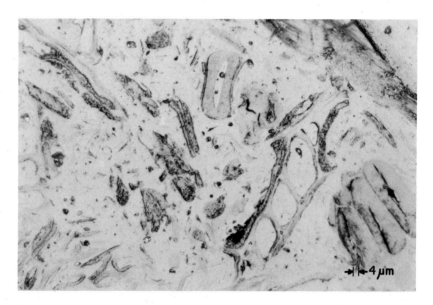

FIGURE 2-19. Wood flour filler in a commercial, phenol-formaldehyde thermoset plastic. Polished section, viewed by vertical, reflected light. By E. Thomas, American Cyanamid Co.

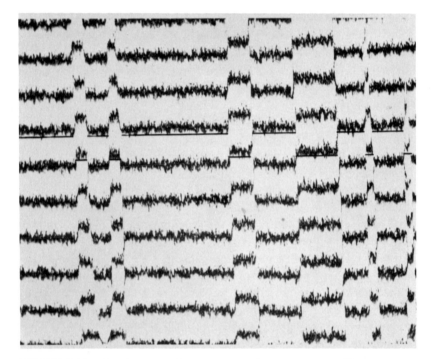

FIGURE 2-20. Interference micrograph of a fracture surface. The reference line indicates the shift in a single fringe as it crosses the boundary of the color areas. By J. P. Berry.[3]

same way that an image is scanned in a TV camera. The reflected and scattered electrons (plus any secondary electrons ejected by the beam) are collected above the specimen, and the weak current is amplified and fed to a TV picture tube or to a synchronized facsimile printer which prints out a "map" of the specimen's surface.[31] The great advantage of the scanning electron microscope is that it pictures surfaces with greater resolution and greater depth of field than the light microscope.

The more common transmission electron microscope can also be used to investigate surfaces, but only in a secondary way. It requires that a *replica* of the surface be made, a replica so thin that the microscope's beam of electrons will go right through it (or at least through the thinnest portions of it). The replica is prepared by applying a very dilute solution of a polymer such as polyvinylalcohol in a highly volatile solvent such as dichloromethane, and then floating off or stripping the replica for examination. Alternatively, a replica can be made by evaporating a thin film of silica onto the surface, and then stripping it off.

The technique of making a replica of the surface to be studied can be used in light microscopy also, and the advantage that light has over electrons is that color can be used for contrast. One way is to use a colored replicating medium (a colored polymer), and to make the replica sufficiently thin so that a colored contour map of the surface is obtained. Another way is to press a fine-grained photographic emulsion (nuclear track plate) against the specimen at 15,000 psi and then to develop it to give a photographic image in which the elevations of the surface are darker than the depressions.

A surface of high optical reflectivity can be examined by means of the interference microscope. The specimen is illuminated vertically through an optical flat plate in contact with it, and colored interference fringes similar to Newton's rings are produced. The order of color corresponds to the contour lines of the familiar topographical map.[32] A direct tracing of topography can also be obtained sometimes, as when a single fiber is drawn under a fragment of razor blade coupled mechanically to a stylus which draws a profile map of the fiber (see the example in Figure 2-9, prepared at the Textile Research Institute).

LEVEL IV. METHODS FOR INVESTIGATING BULK MATERIALS

In the development and quality control of products such as molded plastics, fabrics, fibers, and films, most producers test for properties or behavior, and they analyze for composition and check on condition. It remains the task of the resinographer to fill in the gaps by investigating structure and morphology.

There is much useful information that can be obtained by resinographic techniques: For example, is a fiber crystalline or amorphous?

Or does it have crystalline domains in an amorphous matrix, or vice-versa? Is the fiber's structure oriented, random, or only partly oriented? Can we identify specks or globs that have gotten in by mistake? What are the properties of this or that dispersed phase, or of a tiny sample? We shall consider a few of the techniques that can answer such questions.

The question of crystallinity can be answered by x-ray diffraction of a single fiber (see Figure 2-21). If the answer is intermediate between the extreme examples given in the figure, then the *degree* of crystallinity can be obtained by comparing the intensities of the components of the diffraction pattern on a scanning diffractometer. Such degree of crystallinity is very important in controlling the bulk properties of polyethylene, for example.

Another method for observing crystallinity is to use the polarizing microscope. Here a polarizing device (a **polar**, usually a disc of polarizing film) is placed before the substage condenser of a microscope, and a similar device (called the analyzer) is placed beyond the objective. When one of the polars is rotated so that the planes of polarization are crossed, the field of view appears dark. It remains dark when any **isotropic** medium (a glass slide) or specimen (a plate of sodium chloride cleaved from a salt crystal) is put on the stage, but an **anisotropic** specimen appears bright because it separates the light into two independent beams which are brought together again in the analyzer and, if in phase, produce the brightness. A typical example is given in Figure 2-22, which shows a rosin "**size**" used extensively by the paper industry. This is a milky suspension of rosin particles in an aqueous solution of sodium salts of the rosin acids. The utility of a certain kind of size hinged on the degree of crystallinity and the kind of crystals that developed upon incubation at a certain temperature for a certain time. Under some conditions undesirable (insoluble) acicular crystals developed, while under other conditions harmless soluble spherulites were formed. The difference could not be discerned by running the usual tests such as density, viscosity, and pH.

The physical properties of small samples or of tiny domains in a bulk sample can be determined on a microscopical hot stage[33,34] *between crossed* **polars**. Melting points and other phase transitions show up as a change in polarization colors at any abrupt thermal end point.

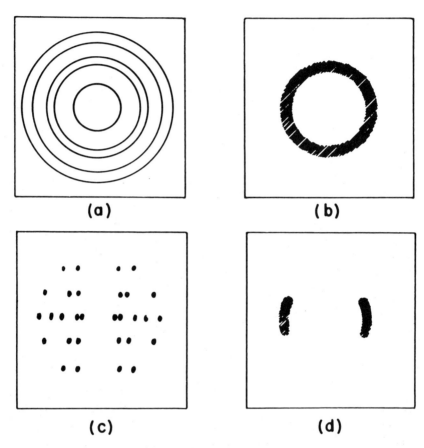

FIGURE 2-21. Diagrammatic drawings of characteristic x-ray diffraction patterns. (a) Unoriented crystalline resin (as a fiber). (b) Unoriented amorphous resin (as a fiber). (c) Oriented crystalline resin (as a fiber). (d) Oriented amorphous resin (as a fiber). Photo courtesy of L.A. Siegel, leader of X-ray Group, Central Research Div., American Cyanamid Co.

0.1mm

FIGURE 2-22. Crystalline phase in a rosin size emulsion with a high free-rosin
content between crossed polars (200 ×).

Even an array of phases can be observed simultaneously, and their joint behavior determined as the temperature is changed.

The microscopical study of small samples and tiny domains does not take the place of more conventional testing of bulk samples, of course. Physical tests and studies of behavior should be done as much as possible on conventional-sized samples, under standard procedures, and with standard equipment. Only in this way can comparable results be obtained on different samples by different people. The preferred standards are those of the ASTM, and especially those sponsored by the national Committee D-20 on Plastics. Such tests include the measurement of mechanical, thermal, electrical, and optical properties, and the study of time-dependent behavior such as the weathering of plastics. There are ASTM tests also available for some chemical analyses and for the evaluation of environmental condition, appearance, and color change.

Bulk testing often includes a search for trace elements, particularly those that are catalytic or exert an inordinately large effect. Often these are inadvertently present, but must be monitored. Sometimes trace elements are added by a manufacturer to help identify his product, even though mixed or blended with other products. A very old method of detecting such trace elements is atomic absorption spectrophotometry, recently revived and refined to an elegant degree as a single instrument capable of detecting and estimating three-fourths of the chemical elements.[35] A few of the elements (but only a few) can also be detected in small amounts by means of their natural radioactivity. Where the end use permits it, other elements can be followed even in trace amounts by incorporating radioactive tracer isotopes in the product, in which case the standard counting equipment and procedures apply.[36]

Differential thermal analysis[37] is a destructive method for measuring the thermal stability and the mode of thermal decomposition of an organic polymer or plastic. The apparatus consists of a closely controlled and programmed furnace with associated thermometry circuits and recorder that continuously measures the temperature difference between a sample of material and a reference (say of porcelain or refractory alloy) as the two are heated simultaneously in the furnace at a pre-set rate. Any phase change that occurs in the sample while it is being heated will cause it to absorb or evolve heat, leading to a temperature differential between it and the reference block. Similarly, any decomposition reaction which occurs in the controlled inert atmosphere must be endothermic (usually) or exothermic (rarely), and these changes will lead to negative and positive differentials, respectively. The temperature differential is plotted automatically against time, giving a sequential record of all the changes that take place in the sample[5] (see Figure 2-14).

A related but distinctly different method of investigation is *thermogravimetric analysis* (TGA),[38] in which the *weight* of a sample is recorded continuously while the sample is being heated at a controlled rate in a controlled atmosphere. If it is heated in nitrogen, the onset of each stage of thermal decomposition is recorded as a loss of weight; if heated in air or oxygen, the onset of each stage of oxidation is recorded. The temperature is recorded on the same chart. Since the heating is done in an enclosed cavity, it is possible to collect and analyze the gases or vapors given off during the decomposition. Better still, the effluent can be conducted directly into a gas-phase chromatograph to identify the decomposition products. Figure 2-13 identifies the depolymerization products obtained from a sample of plastic film, and also the moisture and volatile plasticizers which come off earlier. A series of measurements at different rates of heating can give additional information. Colorimetric determination of the **glass transition** temperatures of polymers is especially useful.[39]

Color may seem to be less fundamental than strength or durability as a bulk property, but often the color or the change of color is a sensitive indication of some condition or change which may be important. Color is produced by electronic transitions which occur with the absorption of particular wavelengths of light (the subsequent emission of the absorbed energy occurs at much longer wavelengths, usually in the heat range). Although color names and color systems (Munsell, etc.) are still used in a qualitative way, the definitive and quantitative description of color is made by a **recording spectrophotometer.** The resulting record indicates percent transmission of absorption as a continuous function of wavelength or frequency, and such a record defines the physical color of the sample. To correlate this with the visual impression requires a set of guidelines, usually those of the International Commission on Illumination known as the tristimulus system of colorimetry. Such color difference units are used in Figure 2-10, which illustrates the change in color of samples of polyester exposed outdoors for a period of months.

SUMMARY

This chapter is the longest and in some ways the most important in this book. It concerns (1) the methods used to study resins, polymers, plastics, and the myriad products made from them, and (2) the recording of the resultant information in useful form. The interpretation of the recorded information will be taken up sequentially in the following chapters.

Records, whether they be photographs, recorder tracings, drawings, graphs, or whatever, must be organized according to some scheme of retrieval. Punch cards are recommended for small personal files, and encoded computer input for larger institutional use. The choice of system is largely a personal matter, but the system should be organized and implemented *before* samples and data accumulate.

Resinographic investigations can be carried out at any one or more (or all) of the four levels described in Chapter 1. At the molecular level, infrared spectrophotometry (IR) is the most common and convenient method for determining content and structure. The next most informative method, especially when hydrogen is a principal constituent, is nuclear magnetic resonance (NMR), which is much more complicated and expensive but yields unique information about structure and bonding in polymers. Electron spin resonance (ESR) tells about free radicals and the efficacy of stabilizers in plastics. Ultraviolet (UV) spectroscopy tells about excitation of molecules, especially of unsaturated groups and conjugated systems. Closely related (and sometimes possible with the same instrument) is spectrophotometry in the visible region, which defines color and measures precisely the changes that take place on exposure or aging. A very different kind of information is given by mass spectrometry (MS), which identifies the fragments of decomposed or pyrolyzed polymer by measuring their actual "weights" in atomic mass units. Molecular weights of undecomposed polymers can, of course, be determined by classical colligative methods and by light scattering or equilibrium centrifugation, but in each case a particular kind of average molecular weight is obtained, and it is necessary to understand the differences and relation between these averages. For all the instrumental methods of investigation, a brief account of the principle of operation is given, followed by limitations and advantages and then by references to detailed treatment and to compilations of spectra or data.

At the level of *phases,* materials are studied most conveniently by microscopy. Several kinds of microscopes (each with its own applicability and limitations) can be applied to the study of constituent phases and the determination of their melting points, phase-transition temperatures, hardness, rate, and degree of crystallization, and so on. In addition, x-ray diffraction techniques may be used to determine crystal structure, even using x-rays that originate in a transmission electron microscope (TEM).

The *surfaces* of polymers, plastics, and fibers are best studied by means of the scanning electron microscope (SEM), which bombards the surface systematically with electrons and picks up the reflected or scattered ones. Surfaces can also be investigated by the indirect method of making a thin replica, stripping it off, and then examining the contours of the replica by light or electron microscopy. Depending on the degree of irregularity, surfaces can also be studied by methods ranging from direct tracing to interferometry. References to all are given.

Lastly, at the engineering level, the bulk properties of a material can be studied by chemical analysis of its constituents, by determining strength and durability of standardized samples by prescribed ASTM methods, and by actual- or accelerated-use testing. The overall degree of crystallinity can be determined by x-ray diffraction, or more conveniently by observing the characteristic behavior of anisotropic crystals by means of the polarizing microscope. The identification of intentional additives, accidental inclusions, and alteration products can also be done microscopically. In addition, the bulk behavior on heating or oxidation can be studied by differential thermal analysis (DTA), which records the changes in heat content as phase changes or chemical reactions take place, or by thermogravimetric analysis (TGA), in which the sample is heated in a desired atmosphere and its changes in weight are recorded continuously. The volatile products can be examined simultaneously by vapor chromatography or mass spectrometry.

Throughout this chapter the reader has encountered many new words, and he should not let any of them go unchallenged or not understood. A thorough glossary appears on pages 165–184, and the references (plus the following Suggestions for Further Reading) should be consulted freely.

SUGGESTIONS FOR FURTHER READING

J. A. Moore and D. L. Dalrymple, *Experimental Methods in Organic Chemistry*, W. B. Saunders Co., Philadelphia, Pa. 19105 (1971).

Polymer Science and Technology: A Symposia Series, Plenum Publishing Corp., New York, N.Y. 10011. Vol. 1: *Structure and Properties of Polymer Films* (R. W. Lenz and R. S. Stein, eds.) (1973); Vol. 2: *Water-Soluble Polymers* (N. M. Bikales, ed.) (1973); Vol. 3: *Polymers and Ecological Problems* (J. E. Guillet, ed.) (1973).

D. G. Peters and E. Wehry, *Instrumental Analysis*, W. B. Saunders Co., Philadelphia, Pa. 19105 (1975).

See Vincent J. Schaefer, Surface replicas containing dye for use in the light microscope, in *Metal Progress* **44**, 72–74 (1943), and the many other contributions of this ingenious experimenter to the study of crystals and surfaces. It was Schaefer who first induced precipitation by seeding clouds with dry ice.

Resinographic Methods (T. G. Rochow, ed.), ASTM Special Technical Publication 348, a symposium of 16 papers distributed over the four levels of organization from molecule to material, American Society for Testing and Materials, Philadelphia, Pa. 19103 (1964).

REFERENCES

1. A. F. Kirkpatrick, An application of punch cards, filing of optical properties of crystals, *Analytical Chem.* **20**, 847–849 (1948).
2. T. G. Rochow, Resinography of high polymers, *Analytical Chem.*, **33**, 1810–1816 (1961).
3. *Morphology of Polymers* (T. G. Rochow, ed.), Interscience Div., John Wiley and Sons, Inc., New York, N.Y. 10016 (1963); also published in *J. Polymer Sci., Part C*, Symposia, **3**, iii–164 (1963).
4. *Resinographic Methods* (T. G. Rochow, ed.), ASTM Special Technical Publication 348, American Society for Testing and Materials, Philadelphia, Pa. 19103 (1964).
5. *Thermal Analysis of High Polymers* (B. Ke, ed.), *J. Polymer Sci., Part C, 6*, 1–214 (1964).

6. N. L. Alpert, W. E. Keiser, and H. A. Szymanski, *IR-Theory and Practice of Infrared Spectroscopy*, Plenum Publishing Corp., New York, N.Y. 10011 (1970); also published in paperback by Plenum/Rosetta (1973).
7. H. A. Szymanski, *Interpreted Infrared Spectra*, Plenum Publishing Corp., New York, N.Y. 10011 (Vol. 1, 1964; Vol. 2, 1966; Vol. 3, 1967).
8. H. A. Szymanski and R. E. Erickson, *Infrared Handbook*, revised ed., Plenum Publishing Corp., New York, N.Y. 10011 (1970).
9. D. O. Hummel, *Infrared Analysis of Polymers, Resins, and Additives; An Atlas*, Vol. 1, Part 1, Interscience Div., John Wiley and Sons, Inc., New York, N.Y. 10016 (1968).
10. A. D. Cross and R. A. Jones, *Introduction to Practical Infra-Red Spectroscopy*, 3rd ed., Plenum Publishing Corp., New York, N.Y. 10011 (1969).
11. J. A. Moore and D. L. Dalrymple, *NMR Spectra of Unknowns*, W. B. Saunders Co., Philadelphia, Pa. 19105 (1971).
12. F. A. Bovey, *High Resolution NMR of Macromolecules*, Academic Press, New York, N.Y. 10003 (1972).
13. R. H. Bible, *Interpretation of NMR Spectra*, Plenum Publishing Corp., New York, N.Y. 10011 (1965); *Guide to the NMR Empirical Method*, Plenum Publishing Corp., New York, N.Y. 10011 (1967).
14. H. A. Szymanski and R. E. Yelin, *NMR Band Handbook*, Plenum Publishing Corp., New York, N.Y. 10011 (1968).
15. *Formula Index to NMR Literature Data* (M. G. Howell, A. S. Kende, and J. S. Webb, eds.), Plenum Publishing Corp., New York, N.Y. 10011 (1966).
16. I. Y. Slonim and A. N. Lyubimov, *The NMR of Polymers*, Plenum Publishing Corp., New York, N.Y. 10011 (1970) (translated from the Russian).
17. J. E. Wertz, Nuclear and electronic spin magnetic resonance, *Chem. Reviews* 55, 828–955 (1955). For a general discussion, see *McGraw-Hill Encyclopedia of Science and Technology*, 3rd ed., McGraw-Hill Book Co., New York, N.Y. 10020 (1971).
18. *UV Atlas of Organic Compounds* (Photoelectric Spectrometry Group, England, and Institut für Spektrochemie und Angewandt Spektroskopie, Germany, eds.), Plenum Publishing Corp., New York, N.Y. 10011 (Vol. 1, 1966; Vol. 2, 1967; Vol. 3, 1967; Vol. 4, 1968; Vol. 5, 1971).
19. J. R. Edisbury, *Practical Hints on Absorption Spectrometry (Ultraviolet and Visible)*, Plenum Publishing Corp., New York, N.Y. 10011 (1967).
20. V. S. Fikhtegol'ts, R. V. Zolotareva, and Yu. L'vov, *Ultraviolet Spectra of Elastomers and Rubber Chemicals*, Plenum Publishing Corp., New York, N.Y. 10011 (1966) (translated from the Russian).
21. K. Hirayama, *Handbook of Ultraviolet and Visible Spectra of Organic Compounds*, Plenum Publishing Corp., New York, N.Y. 10011 (1967).
22. *A Handbook on Mass Spectroscopy*, National Research Council Publication No. 311, Washington, D.C. 20418 (1954); H. E. Duckworth, *Mass Spectroscopy*, Cambridge Monographs on Physics Series, Cambridge University Press, New York, N.Y. 10022 (1958).

23. *Modern Aspects of Mass Spectroscopy* (R. I. Reed, ed.), Plenum Publishing Corp., New York, N.Y. 10011 (1968).
24. *Mass Spectrometry and NMR Spectroscopy in Pesticide Chemistry* (R. Hague and F. J. Biros, eds.), Plenum Publishing Corp., New York, N.Y. 10011 (1974).
25. *Gel Permeation Chromatography* (K. H. Altgelt and L. Segal, eds.), A.C.S. Symposium, Houston, 1970, M. Dekker, New York, N.Y. 10016 (1971).
26. R. M. Kimmel and R. D. Andrews, Birefringence effects in acrylonitrile polymers. II. The nature of the 140° C transition, *J. Applied Physics* **36,** 3063–3071 (1965).
27. W. C. McCrone, *Fusion Methods in Chemical Microscopy*, Interscience Div., John Wiley and Sons, Inc., New York, N.Y. 10016 (1957).
28. A. V. Signeur, *Guide to Gas Chromatography Literature*, Plenum Publishing Corp., New York, N.Y. 10011 (Vol. 1, 1964; Vol. 2, 1967; Vol. 3,1974).
29. S. G. Perry, R. Amos, and P. I. Brewer, *Practical Liquid Chromatography*, Plenum Publishing Corp., New York, N.Y. 10011 (1972); also available in paperback from Plenum/Rosetta (1973).
30. E. P. Bertin, *Principles and Practice of X-Ray Spectrometric Analysis,* 2nd ed., Plenum Publishing Corp., New York, N.Y. 10011, (1975).
31. *McGraw-Hill Encyclopedia of Science and Technology,* 3rd ed., McGraw-Hill Book Co., New York, N.Y. 10020, (1971).
32. S. Tolansky, Microstructures of Surfaces using Interferometry, Edward Anold, London, Great Britain (1968).
33. T. G. Rochow and R. J. Bates, A microscopical automated microdynamometer and microtension tester, *ASTM Materials Research and Standards* **12**, No. 4, 27–30 (1972).
34. E. M. Chamot and C. W. Mason, *Handbook of Chemical Microscopy*, Vol. 1, *Physical Methods*, 3rd ed., John Wiley and Sons, Inc., New York, N.Y. 10016 (1958).
35. R. Lockyer, Atomic absorption spectroscopy, in *Advances in Analytical Chemistry and Instrumentation,* Vol. 3, Interscience Div., John Wiley and Sons, Inc., New York, N.Y. (1964); J. Dean, *Flame Photometry*, McGraw-Hill Book Company New York, N.Y. (1960); M. L. Parsons and P. M. Elfresh, *Flame Spectroscopy: Atlas of Spectral Lines*, Plenum Publishing Corp., New York, N.Y. 10011 (1971).
36. G. Friedlander, J. W. Kennedy, and J. M. Miller, *Nuclear and Radiochemistry,* 2nd ed. (Chaps. 5, 6, and 7), John Wiley and Sons, Inc., New York, N.Y. 10016 (1964); *Radiochemical Methods in Analysis* (D. O. Coomber, ed.), Plenum Publishing Corp., New York, N.Y. 10011 (1974).
37. H. Ferrari, in *Thermal Analysis* (K. F. Schwenker and P. D. Garn, eds.), Vol. 1, p. 48, Academic Press, New York, N.Y. 10003 (1969).
38. *Thermal Analysis of High Polymers* (B. Ke, ed.), Interscience Div., John Wiley and Sons, Inc., New York, N.Y. 10016 (1964).
39. W. P. Brennan, Differential scanning colorimetry: a method for polymer quality control, *American Laboratory*, pp. 75–81, (Feb. 1975).

RESINS AND POLYMERS
AT THE MOLECULAR LEVEL

In this chapter we shall consider the structure, morphology, composition, condition, properties, and behavior of the molecules which make up resins, polymers, and elastomers. We shall concentrate on the intrinsic attributes of the pure chemical compounds, whether they be man-made or isolated from natural products by purification procedures. Such attributes are for scientific, rather than practical, purposes; they satisfy natural curiosity and lead to a basic understanding of the material. The information is also to be used in conjunction with technological and practical information gathered on Levels II, III, and IV.

There are many questions to be answered about *structure* at the molecular or macromolecular level. If macromolecular, are the substances homopolymers or copolymers? Are they chainlike, or branched, or cyclic, or all three? What isomeric configurations (*cis* or *trans*, isotactic,* syndiotactic, or atactic) are present? Are the molecules inherently capable of organizing themselves into crystals, or not? Are the specific structural formulas known? All these structural

---

*The term *atactic* refers to an irregular or random arrangement of side groups about the polymer chain, leading usually to an amorphous polymer. *Isotactic* polymers have a stereoregular structure with all the side groups on one side of the chain (as a result of polymer growth with stereospecific catalysts). *Syndiotactic* polymers also are stereoregular, with pendant groups on alternating sides of the chain. They and isotactic polymers tend to crystallize. See Glossary.

aspects depend on the chemical composition, and taken together they determine such inherent properties as chemical reactivity and potentially thermosetting or thermoplastic behavior.

The questions of *morphology* are almost as numerous and fundamental. What is the conformation of the molecules (that is, what is the degree of twist or angle of the spiral, if any)? Are the "linear" molecules coiled or uncoiled? What is the degree of polymerization? What is the average molecular weight? What is the range of molecular weights, and the distribution of them? What is the degree of crystallizability and of crystallinity, in relation to molecular size and shape? These and other morphological considerations, together with the environmental treatment involved, will affect the *rates* of reaction, dissolution, melting, thermosetting, and plastic deformation.

These concepts on the molecular level are outlined in Table 3.1 for some types of natural, modified, or man-made resins, polymers, rubbers, gums, and starches, and are illustrated by the following examples.

Natural resins of recent origin include rosin, dammar, copal, acaroid, elemi, mastic, sandarac, and shellac. These are physical mixtures (mutual solutions) of single structural units, chiefly complicated organic acids and esters. Rosin, for instance, is composed chiefly of some seven kinds of tricyclic acids, most of them having the empirical formula $C_{19}H_{29}COOH$ with a molecular weight of 302. The acid content depends upon the treatment used to obtain the rosin. "Gum" rosin from the exudate of pine trees and "wood" rosin (from pine stumps) are 90% rosin acids, and "tall oil" rosin (by-product of making kraft paper from pine trees) is 95%.[1] The unsaponifiable parts are known as resenes. The differences among three of the principal stereoisomeric, condensed, tricyclic acids are shown by their structural formulas, which are given in Figure 3-1. The three stereoisomers differ in the position of the double bonds and in the consequent degree of rotation of the plane of polarized light. Indeed, one kind of pimaric acid has the property of rotating the plane of polarization to the right (*d*) and the other to the left (*l*), because of their respective molecular configurations. Hence **polarimetry** is an important method of differentiation.

Chemical methods are of fundamental value to determine the percentage of saponifiable molecules. The acid reactivity and alcohol solubility of rosin are two of the very practical properties. The sodium soap is very soluble in water, but the aluminum soap is not (a behavior that is very important in sizing paper). The cobalt, manganese, and lead soaps are important catalysts ("driers") for linseed and similar drying oils.

The other natural resins of recent vegetable origin also are mixtures of unit molecules, chiefly organic acids. Shellac and other lacs are of recent insect origin and are chiefly mixtures of unimeric interesters of various polyhydroxy carboxylic acids.

Natural resins of recent origin often are modified industrially to increase molecular stability and to reduce crystallizability. Rosin, for instance, is made more stable in air and water and less crystallizable by dimerization or tetramerization brought about by heating, or it is stabilized by hydrogenation. It can not be polymerized, but it can be molecularly stabilized by esterifying it with a polyhydric alcohol such as glycol or glycerin to make oil-soluble ester gum for use in varnishes and paints.

Medium-fossilized resins, such as some of the copals, are naturally polymerized to the extent that they are not alcohol-soluble but are soluble in drying oils such as linseed oil, so as to make varnish. Highly fossilized resins such as Zanzibar copal, kauri, East India, and amber are insoluble in linseed oil in the cold, and need to be "run" (heated), presumably to reduce their degree of polymerization so as to make them soluble.

The gray-brown viscous sap from the *Rhus vernicifera* (varnish tree) darkens in moist air and forms a tough, protective film known as Japanese or Chinese lacquer. The mechanism of hardening involves the condensation of phenolic compounds, the chief one being urushiol, $C_6H_3(OH)_2C_{15}H_{27}$. Acaroid resin (called "gum accroides" by varnish makers) and cashew oil also contain phenolic substances.

Linseed oil, the most important natural drying oil, is oxidized in air at the three double bonds of linolenic acid and at the two double bonds of linoleic acid, both of which are present along with palmitic, myristic, oleic, and isolinolenic acids, as mixed triglycerides. Thus linseed oil, like all fatty oils and fats, is a variable mixture of molecules. Tung (China wood) oil is another important drying oil. Fish oils, primarily menhadin oil, contain glycerides of cluponodonic acid, which has five double bonds. Linseed and other drying oils are modified (partially oxidized) by boiling (heat-bodying) them.

TABLE 3-1.

Resin, etc.	Structure (kind)	Morphology (extent)
Natural, recent	Fused cyclic acids	Acidity (acid no.)
Natural, modified	Hydrogenated; dimerized or esterified	Extent modified
Natural, fossilized	Polymers, obscure	Degree of polymerization
Modified, fossilized	Lower polymers	Degree of depolymerization
Natural polymers	Cellulose; proteins	Very high molecular weight
Regenerated polymers	Cellulose; proteins	Lower molecular weight
Modified polymers	Cellulose acetate; protein aldehyde	Lower molecular weight
Synthetic condensates	$-\phi OH$; $(-RNH_2)$; R_2CHO	Degree of condensation
Addition polymers	Config. (*cis* or *trans*)	Conformation; molecular weight
Rubber; elastomers	Like *cis*-isoprene?	Cross-link; molecular weight
Gums	Complex sugar acids	Unit molecules (?)
Starches	Complex carbohydrates	Polymeric
Starches modified	Depolymerized, etc.	Degree degeneration

Typical Materials in Terms of Their Constituent Molecules (Level I)

Composition	Condition	Properties	Behavior
Various unit molecules	As extracted, purified	Reactive; soluble in alcohol	Unstable
Various unit molecules	Heated; hydrogenated; esterified	Insoluble in alcohol; soluble in oils	More stable
Still obscure	Prehistoric	Insoluble in alcohol; soluble in oils	Most stable
Still obscure	Heated ("run")	Soluble in oils	More practical
Empirical	Separated, sorted, cleaned	Insoluble in most solvents	Natural
Empirical	Spun, cast, or molded	Insoluble in most solvents	Dependent on conditions
As modified	Spun, cast, or molded	Soluble in most solvents	Dependent on conditions
Specific	Cast, hot-pressed	Insoluble in most solvents	Stable in air, H_2O
Homogeneous vs. copolymer	Cast; ejected or extruded; pressed	Soluble in some solvents	Stable in air, H_2O
Homogeneous vs. copolymer	Cast; ejected or extruded; pressed	Rubbery at room temperature	Stable in air; solvent
$C_{23}H_{38}O_{22}$, etc.	Separate, sorted, cleaned	Soluble in H_2O	Disposable, edible
$(C_6H_{10}O_5)_n$	Separated; purified	Insoluble in H_2O; paste	Varies with source
Lower degree of polymerization	Heated, acidified or fermented	More soluble in H_2O	Varies with source

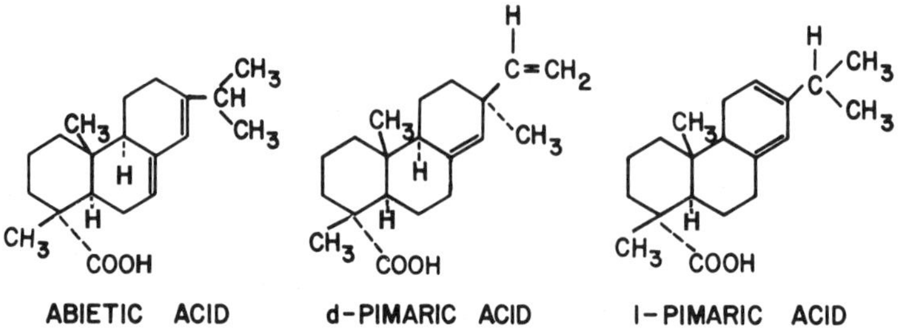

ABIETIC ACID   d-PIMARIC ACID   l-PIMARIC ACID

FIGURE 3-1.   Structural formulas of three rosin acids.

Raw rubber can be vulcanized (cross-linked at its double bonds) with sulfur or sulfides. Cellulose from wood or cotton is chemically treated to render it soluble in water, so that it may be regenerated into cellulose as viscose, rayon, Bemberg, or other fibers. It may also be converted to a cellulose ester, such as the acetate, or it may be partially regenerated before converting into fibers.

Molecular differences of *extent* are exemplified by practically all polymeric materials, natural or man-made. Except in rare instances of monodispersion, high polymers are mixtures of molecules possessing various degrees (n) of polymerization, and, consequently, of various molecular weights. Therefore, the averages and distributions of molecular weights are morphological characterizations on Level I. Different processes of polymerization may produce different average degrees of polymerization. For instance, much larger macromolecules of poly(methyl methacrylate) are produced by casting the monomer than by bulk-polymerizing it. Of course, various degrees of polymerization may be obtained by the same process. The degree of polymerization is especially important in polymers obtained by condensation, such as by reacting a phenol with an aldehyde such as formaldehyde. If the reaction is carried to completion, the resulting product is insoluble in practically every physical solvent. But Baekeland discovered around 1900 that if he interrupted the reaction before completion, he could obtain a resin which could be compression molded, or dissolved in an alcohol and used as a solution.

Amine-aldehyde condensates, such as Melmac® melamine-formaldehyde resin, are soluble in water if they are in the early stages of condensation. If completely cured (as in Melmac® tableware) the resin is insoluble. But if the monomeric melamine is alkylated before partial reaction with formaldehyde, the resulting uncured resin is soluble in paint solvents. Of course, other amines, such as urea or thiourea, may be substituted for melamine with corresponding changes in molecular properties and behavior.

Addition polymers are varied in composition by varying the monomer, thus producing the various kinds of homopolymers. Intermediate variations are prepared by the co-polymerization of two or more monomers. Variations in structure are obtained with copolymers by different sequential arrangements of the monomers: random, alternate, block, or graft.[2]

It is well known that the effective *shape* or conformation of long macromolecules is varied by the nature of contiguous solvent molecules.[3] A "good" solvent is one which relaxes or uncoils the molecular chain. A "poor" solvent is one which tightens or coils it. The effects of both kinds of solvents are shown side by side in Figure 3-2, an electron micrograph of polyacrylamide molecules which were dried from solution in a mixture of water and methyl alcohol. Methyl alcohol is a poor solvent, and its presence causes the molecules to coil up into balls, as shown in the upper right-hand corner of Figure 3-2. As the methyl alcohol evaporates preferentially, a concentration of water is reached wherein the molecules are relaxed into long stringlike shapes, as shown in the lower left-hand corner of the electron micrograph. The presence of both coiled and uncoiled molecules in the dry state indicates that the different effects of contiguous molecules persist even when one is dealing with only the solute molecules. We conclude that there may be corresponding differences in the morphology of commercial high-polymeric molecules, depending on whether the monomer polymerizes in a solvent, or as an emulsion, or by direct casting of the catalyzed liquid monomer.

Rubber is a generic term given to any composition that behaves and feels something like natural rubber. The word embraces not only the polymeric substance which gives rise to the rubbery or elastic property, but also the various fillers, reinforcing agents, plasticizers, extenders, pigments, antioxidants, and vulcanizing agents with which that polymer is formulated. For this reason it is somewhat more exact to refer to the rubbery polymer itself as an *elastomer*, and to think of it

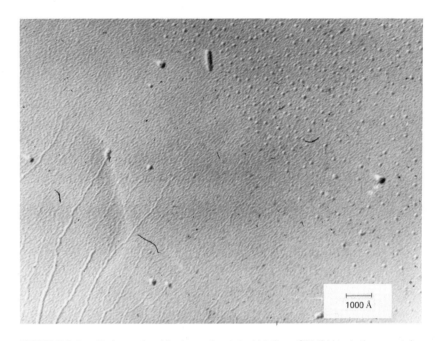

FIGURE 3-2.   Polyacrylamide from droplet of H_2O + CH_3OH solution containing NH_4Cl electrolyte. Dried polymer shadowed with uranium. Center of dried
droplet is outside of lower left-hand corner of picture. Lower left: inner sector of
droplet; branched and complicated filaments. Intermediate: zone of thin, single
filaments, like uncoiled macromolecules. Upper right: outermost zone; globules
such as is expected of coiled macromolecules.

(at least for the present) in terms of polymer chemistry, quite apart
from all the additives that may go into the final product later.

Seen this way, "rubber" refers to a state of matter (the elastomeric state), rather than to any particular chemical composition or
kind of substance. Thus there are entirely inorganic rubbers, such as
polyphosphonitrilic chloride; there are organometallic rubbers, such as
silicone rubber; and there are many dissimilar organic rubbers of the
diene, copolymer, and polysulfide types. Almost any resinous thermoplastic polymer enters upon a rubbery stage when heated
sufficiently,* and conversely, any elastomer loses its elasticity and be-

*For example, ordinary polystyrene becomes rubbery in the range from 100°C to its
 melting point.

comes stiff when cooled sufficiently. The temperature at which that occurs is called the *glass transition temperature*,* which is $+ 100°C$ for polystyrene, $- 73°C$ for natural rubber, and $- 109°C$ for silicone rubber. The addition of various solvating agents or plasticizers to a stiff, glassy polymer may serve to extend the rubbery range downward, as when tricresyl phosphate is milled with polyvinyl chloride to make a popular rubbery wire insulation. Similarly, the familiar polyethylene is a flexible plastic at room temperature, and so is polypropylene, but a particular copolymer of ethylene and propylene (in which they solvate each other) is a serviceable and chemically stable elastomer. Both effects are seen to be an extension of the preceding discussion of the effects of contiguous solvents on polymer molecules.

Natural rubber is essentially polyisoprene in the *cis* configuration of the unsaturated, asymmetric carbon atoms[4]:

$$-CH_2-CH_2-\underset{\underset{CH_3}{|}}{C}=\underset{\underset{H}{|}}{C}-CH_2-CH_2-\overset{\overset{CH_3}{|}}{C}=\overset{\overset{H}{|}}{C}-CH_2-CH_2-$$

$$\longleftarrow \text{—————— identity period 9.13AU ——————} \longrightarrow$$

Natural rubber

Here the repeating unit of natural rubber (*Hevea brasiliensis* or caoutchouc) is found to be 9.13 AU by x-ray diffraction, and this long *cis* configuration allows the long molecules to coil and uncoil with relative freedom. Compare the molecular structure of gutta-percha (*Palaquium gutta*), the natural polymer used for the outer casings of golf balls:

$$-CH_2-CH_2-\underset{\underset{CH_3}{|}}{C}=\overset{\overset{H}{|}}{C}-CH_2-CH_2-\underset{\underset{CH_3}{|}}{C}=\overset{\overset{H}{|}}{C}-CH_2-CH_2-$$

$$\longleftarrow \text{identity period 5.04 AU} \longrightarrow$$

Gutta-percha

*Glass transition temperature, T_g, is generally taken as the midpoint of a range covering the observed transition.

Here there is a *trans* configuration about the double bonds of polyisoprene, and this, together with the much shorter repetitive unit of
structure, encourages the molecules to align rather than to coil in loose
spirals.

Natural rubber has advantages which have prevented its complete
displacement by man-made elastomers of other structure. It is highly
elastic, and there is little internal energy absorption as it stretches and
relaxes because the heating is balanced by a consequent cooling.* It
vulcanizes readily at the double bonds to give a strong cross-linked
structure. These advantages are retained and intensified in synthetic
polyisoprene, in which isoprene made from petroleum is polymerized
with a stereospecific catalyst to give pure *cis* polymer.

Natural rubber also has disadvantages. Being unsaturated, it
reacts slowly with oxygen (and rapidly with ozone) to shorten its macromolecules (by scission) and becomes a cracked, inelastic, perhaps
sticky mass. Being a hydrocarbon, natural rubber also burns easily and
dissolves in (or is swelled by) hydrocarbon solvents like gasoline. The
first fault is avoided by man-made elastomers which have no unsatura-
tion, such as polyisobutene and silicone rubber. The second fault
is avoided or diminished by elastomers of different chemical
composition, not so soluble in hydrocarbons, such as polysulfide
rubber, acrylonitrile copolymers, and silicone rubber. The different
compositions of these are readily determined by analysis or by simple
burning of a small sample: polysulfide rubber (Thiokol®) gives a strong
odor of sulfur dioxide, nitrile rubber (Buna® and ABS) gives an odor of
burning nitrogenous material (reminiscent of burning hair or feathers),
and silicone rubber gives little odor, almost no flame, and a white
smoke of silicon dioxide. The various types of elastomer can also be
recognized by refractive index, specific gravity, and infrared spectra
after the polymer is isolated from fillers and other extraneous matter by
extracting with a suitable solvent and recovering the dissolved polymer
by evaporation.

*The heating effect of stretching and the cooling effect of relaxation can easily be dem-
onstrated with a rubber band. Stretch a stout rubber band quickly and hold it to the lips
or check at once; it feels warm. Wave it in the air while fully stretched, and then release
it quickly: it feels cold.

SUMMARY

It is at the molecular level that resins, gums, starches, and the various natural and man-made polymers and elastomers differ most markedly. This comes about because the terms *gum, resin, rubber,* etc., refer to states of matter, not to specific compositions or to classes of chemical compounds. Anything that looks or acts like rosin is resinous, and anything that behaves like natural rubber is rubbery, regardless of composition. This situation results in a vast array of different chemical substances being employed as resins, gums, and elastomers in materials of commerce. Only a few of the most important of these substances could be considered in this chapter, for reasons of space. A complete catalog of them would take many volumes.*

Natural resins include rosin, copal, shellac, and many others, all not polymers in the usual sense of the term, but consisting instead of mixtures of complex acids and esters. Amber is a fossilized natural resin of vegetable origin, often containing pollen, parts of insects, and other preserved matter from a bygone age. Natural resins can be modified by heating or hydrogenating them, and they often are dissolved in drying oils to make varnishes.

Man-made resins may be addition homopolymers, copolymers of two or more monomers, or condensation polymers. If saturated, they can only be modified by adding plasticizers or by milling together two or more compatible (mutually soluble) resins. The resin can then become the basis of what is commonly called a *plastic* by mixing it with dyes, pigments, fillers, reinforcing agents, antioxidants, and other additives to achieve the desired properties. If the resin is unsaturated, it may be cross-linked or vulcanized by some agent which reacts with its double bonds, thereby tying together many of the large molecules to achieve greater strength or rigidity.

Almost all polymers change from a rigid or brittle state to a more elastic or rubbery state at some characteristic temperature called the

---

*See, for example, the nearly 20 volumes of the *Encyclopedia of Polymer Science* (H. F. Mark, editor), Interscience Div., John Wiley and Sons, Inc., New York, N.Y. which took over ten years to compile.

glass transition temperature, indicated by the symbol T$_g$. If a polymer has a T$_g$ in the neighborhood of room temperature, and especially if it maintains its rubbery condition over a wide range of temperatures, it is called an elastomer. Elastomers become the basis of useful "rubbers," just the way nonelastic polymers become the basis of useful "plastics." That is, the addition of reinforcing agents, coloring agents, antioxidants, and stabilizers makes a commercial material out of the pure homogeneous elastomer. If the elastomer is unsaturated (as in natural rubber), then the final mixture (called a rubber "compound") can be vulcanized by sulfur or some other cross-linking agent, thereby altering its structure at the molecular level.

The various kinds of polymers and elastomers present in a commercial material can be identified by chemical analysis or by the various investigative methods of Chapter 2. An experienced investigator can often classify a material just by its feel, its appearance, its behavior on flexing or stretching, and the odor it produces when burned with a match. The odor and the black smoke of burning natural rubber are characteristic; the sulfur-rich elastomers like Thiokol® give forth black smoke and a strong odor of sulfur dioxide, the nitrogenous elastomers give an odor like that of burning protein, and silicone rubber gives very little odor but a whitish smoke.

The structure, morphology, composition, condition, properties and behavior of some common classes of gums, resins, and elastomers are summarized in Table 3-1, which will reward the reader with much more information if he will review it at this point, after reading this summary.

SUGGESTIONS FOR FURTHER READING

T. G. Rochow, Microscopic domains in some synthetic polymers, *J. Applied Polymer Sci.*, **9**, 569–581 (1965).

ACS Symposium on morphology of polymers (T. G. Rochow, ed.,) at Los Angeles, California, April 4–5, 1963, *J. Polymer Sci.*, *Part C* **3**, 1963; reprinted by Interscience Div. of John Wiley and Sons, New York 10016 (1963). There are 16 papers with special attention here on those by V. G. Peck and L. D. Moore on Morphology of large molecules in polyethylene, pp. 9–19; by M. J. Richardson on Molecular weights of amorphous polymers by electron microscopy, pp. 21–29; and by H. P. Wohnsiedler on Morphology of molded melamine–formaldehyde, pp. 77–89.

REFERENCES

1. Article on rosin, in *Encyclopedia of Polymer Science* (H. F. Mark, ed.), Vol. 12, Interscience Div. John Wiley and Sons, Inc., New York, N.Y. 10016 (1970).
2. F. W. Billmeyer, *Textbook of Polymer Science*, 2nd ed., John Wiley and Sons, Inc., New York, N.Y. 10016 (1971).
3. T. G. Rochow, Resinography of high polymers, *Analytical Chem.* **33**, 1810–1816 (1961).
4. L. F. Fieser and Mary Fieser, *Organic Chemistry*, 3rd ed. Reinhold Publishing Co., New York, N.Y. 10001 (1968); H. Beyer, *Organic Chemistry*, 10th ed. (in English), Verlag Harri Deutsch, Frankfurt/Main (1963).

RESINS AND POLYMERS ON LEVEL II: PHASES

Having considered resins, polymers, and elastomers from the stand-point of molecular composition and structure in the previous chapter, we turn now to the differentiation of *phases* with respect to structure, composition, properties, morphology, condition, and behavior. A phase has been defined in Chapter 1 as an apparently homogeneous, tangibly or intangibly separate portion of matter.

If there is only one component A in a resinous or polymeric sys-tem, it is possible to have one amorphous phase* (liquid or solid, or both together at the melting point), or one or more crystalline phases of different crystal structure, or both amorphous and crystalline phases together. If component A is a homopolymer, such as cast poly(methyl methacrylate), the molecules all have the same type of structure but vary somewhat in morphology (that is, in size and possibly in shape).

With *two* component monomer units A and B, as in a copolymer of 95% methyl methacrylate and 5% ethyl acrylate, a single phase or copolymer may be obtained (in this case an easily moldable, internally plasticized copolymer). With a solid polymer A and a minor amount of unimeric liquid B, such as tricresyl phosphate, the liquid B could dis-solve in A and thereby plasticize A. Or, if we have enough liquid B, the solid A could dissolve in B to form a liquid phase.

---

*For present purposes we shall ignore the vapor phase.

If A and B are two solid phases, for example rosin and natural rubber, A may dissolve in B to make a single rubbery phase AB, or B may dissolve in A to make a sticky phase BA.* Such phases are distinguished from their constituent molecules by considering them as a system of components. Similarly, commercial synthetic polymers with two or more components, such as "externally" plasticized polyvinylchloride, are distinguished from chemically pure polymer. A further distinction can later be made at Level IV, when industrial molding compositions may be found to contain other phases such as pigments, fibers, or lamellae mixed into the polymer matrix to form a synergistic multiphase system. It is important to keep these distinctions in mind so as not to confuse the properties on one level with those on another. For example, the refractive index, solubility, and moldability of pure poly(methyl methacrylate) are quite different from those of commercial grades, which contain ethyl acrylate as copolymer.

On Level II, phases must be distinguished clearly even when they do not differ in chemical composition. Crystallized polystyrene, for example, has different properties from those of amorphous polystyrene of identical chemical composition. The amorphous polystyrene may or may not be crystallizable (see Figure 4-1). If nonresinous particles are found to be present, they would be described on the same level by naming the phase: for example, silica would be labeled quartz, diatomaceous earth, or silica glass, but it would not be denoted by the chemical label SiO_2, which does not specify the phase. Level II is especially useful in distinguishing among solid, semisolid, liquid, and elastomeric phases.

Some of the questions encountered on this level are: (1) How many and what kinds of phases are there? (2) What are the *structural* aspects of these phases? (3) Is each phase amorphous, crystalline, or crystallizable? (4) If a phase is found to be amorphous, is it glassy or elastomeric? (5) If the phase is amorphous, has it been oriented by stretching (or by other physical force during fabrication), or is it unoriented and unstrained? (6) If the phase is crystalline (other than isomeric, therefore isotropic), or if it is anisotropic for any other reason (as with most fibers, which are oriented lengthwise), what are

*This is especially likely if B consists of small molecules.

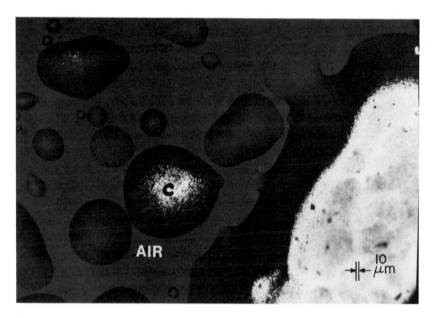

FIGURE 4-1. A mixture of half-crystallizable (isotactic) and half-non-crystallizable (atactic) polystyrene. The small, completely dark droplets have not begun to crystallize at all. In the upper left, a droplet has begun to crystallize. The drop marked C has been considerably crystallized toward its maximum of 50%. On the right, a portion of a large drop has reached its maximum of 50% crystallized.

the optical properties in the principal directions? (7) Which of two principal refractive indices is greater, i.e., what is the optical sign? (8) For any crystalline phase, what are the crystallographic properties? The answers to all these questions will serve to identify the phases and contribute to the understanding of the material.

At Level II there are also some morphological questions: (1) What are the sizes, shapes, and distribution of the phases? (2) What relevant treatment did the sample undergo? (3) What changes did the treatment make? (4) What is the behavior of the sample with time under various influences? Some possible answers to such questions are outlined in Table 4-1.

TABLE 4-1.

Resin, etc.	Structure (kind)	Morphology (extent)
Natural, recent	Amorphous	Moldments, films
Natural, modified	Amorphous	Films, particles
Natural, fossils	Amorphous	As is, or machined
Modified fossils	Amorphous	Films from solution
Natural polymers	Fibrous; cellular	Natural
Regenerated polymers	Oriented	As spun, cast
Modified polymers	Oriented	As spun, cast
Synthetic condensates	Amorphous glass; crystalline	As molded, machined
Addition polymers	Amorphous glass; crystalline	As molded, extruded
Rubber; elastomers	Stereospecific; extensible	As molded, extruded
Gums (e.g., arabic)	Amorphous glass	Nodules
Starches	Naturally crystalline	Natural grains
Starches, modified	Amorphous	Films, powder

Typical Materials in Terms of Their Constituent Phases (Level II)

Composition	Condition	Properties	Behavior
Mixed kinds of molecules	Depends on process	Thermoplastic; soluble	Discolors, devitrifies
Various (soap?)	As result of process	Improved or special	Discolors, crystallizable
Natural, obscure	Natural	Machinable, infusible	Naturally stable
As heated (run)	As modified	Soluble in drying oil	Durable
Natural	Cleaned, refined	Insoluble in organic solvents	Natural
According to process	As regenerated	Insoluble in organic solvents	Predictable
According to process	As modified	Soluble in organic solvents	Predictable
According to cure	Temperature, pressure, tin	Thermoset	Weather, water resistant
Variable	According to poly-merization	Thermoplastic; soluble	Weather, water resistant
Variable	According to poly-merization	Elastomeric	Some affected by air, O_3
Carbohydrates	Cleaned, refined	Colloidal, soluble in water	Adhesive
$(C_6H_{10}O_5)_n$	Naturally crystallized	Swell in water	Stain blue with iodine
Dextrin	As dextrinized	Soluble in water	Stain violet with iodine

ROSIN AND OTHER NATURAL RESINS

Rosin provides a good example with which to illustrate the foregoing considerations. As it comes on the market from several sources, rosin is generally of a single phase, which is amorphous and in the glassy state. Under certain conditions of treatment, however, rosin is partially crystallizable. The crystallizability depends upon the composition, which varies with the process of production, because any change in composition affects the mutual solution of the rosin acids which are present in the amorphous rosin phase.

Rosin also dissolves many other materials, including rubbers. Rosin admixed with natural rubber, for example, is used as a pressure-sensitive adhesive. The highly tacky variety of adhesive is composed of two phases,[1] but they are not pure rubber (A) and pure rosin (B). Rather, both components (on Level I) are in each phase (on Level II). Phase AB is rubber saturated with rosin, while phase BA is rosin saturated with rubber of chiefly lower molecular weight. The AB phase provides pressure-sensitive adhesion, whereas the BA phase provides tackiness. The separate properties and behavior of each phase are on Level II of organization. The structure and morphology of the interfaces between the AB and BA phases and of the surfaces of adhesive and adherents belong on Level III, along with the other corresponding attributes. The practical properties and behavior (life) of the whole laminate of the adhesive and substrate in a standard test or in actual use would be classified with the other corresponding attributes on Level IV.

Natural resinous phases have their own commercial designations. Such identities, if known, should be included in resinographic descriptions. It has been mentioned that "gum" rosin, "wood" rosin, and "tall oil" rosin[2] differ with regard to method of extraction, composition, and properties. Each product of extraction may have various color designations, such as WW ("water white") and WG ("window glass").

Naming species and designating varieties of other tree resins are difficult tasks because they originate in sources which are spread all over the world.[3] Nevertheless, characterization should be attempted on Level II.

The lacs are designated by the shape or size of their particles: flake shellac, stick lac, seed lac, or button lac. Garnet lac has an obvious garnet color. All of the lacs characteristically contain natural plasticizers.

POLYMERS AND ELASTOMERS

For man-made polymers, the differences in classification on Level II versus Level I bear some explanation. On Level I the concern is with chemical species, such as those obtained in the laboratory by synthesis or analysis. On Level II the concern is with practical grades, which may have been modified with dissolved dye or plasticizer, for example, to vary such properties as color, moldability, flexibility, hardness, and so forth. Or the modification may cause a variation of behavior, such as weatherability, stability to light, heat, or electricity, aging, or wear.

There are two kinds of modifiers used as plasticizers: internal and external. Whereas internal modifiers are introduced within molecules on Level I as substituents of or side chains on the monomer, external modifiers are introduced on Level II as separate molecules into the phase to be modified. Successful change from one single phase to another depends on the success in homogeneously incorporating a resin (e.g., rosin), a polymer [e.g., poly(ethyl acrylate)], or a liquid (e.g., tricresyl phosphate) into the host phase. If the host is already in solution, homogeneous incorporation of a modifier in solution requires little more than good stirring. That method is conveniently employed in the formulation of lacquers, varnishes, and paints. The homogenization of melted phases is generally more difficult to accomplish if they are viscous, foamy, or otherwise recalcitrant. While the foregoing statements will seem obvious and straightforward, the literature is often ambiguous because the purity, homogeneity, or technical grade of the plastic is not specified.

Whenever possible, resinographic information about a commercial phase, as received, should contain all the available technical as well as scientific information. For example, the generic name, poly(methyl methacrylate), should be accompanied by the trademark and grade symbol (e.g., Acrylite® M). Provided that the commercial grade has not been modified, the trademark and grade symbol indicate composition (e.g., copolymers) and properties (e.g., refractive index, specific gravity) within the manufacturer's specifications.

A single phase of moldable synthetic thermoplastic polymer may be delivered to the molder in *structural units* such as beads (Figure 4-2), pellets, fragments (Figure 4-3), or spray-dried bubbles (Figure 4-4). Such particles, pellets for example, may have shapes, sizes, and anisotropy which are peculiar to a specific manufacturer and thereby identify him.

More particularly, *fibers* are characterized by the shapes and sizes of cross sections (Figure 4-5), by anisotropy, and seven other optical properties of longitudinal views.[4] If composed of a single phase, fibers may be characterized on Level II. However, these same fibers (or any others) may be in a cut and crimped condition, for example, and then they would also be characterized on Level IV.

FIGURE 4-2. Polystyrene moldable beads. Both specimens were taken by oblique reflected light (3½ ×). Left: exterior; right: cross section.

FIGURE 4-3.   Experimental, moldable resin as fragments (+50–20 mesh) of a
cast product, between crossed polars to show strain anisotropy.

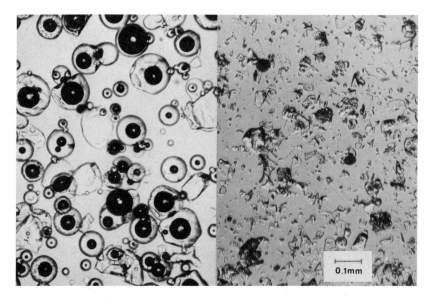

FIGURE 4-4. Spray-dried urea resin. Transmitted light was used. Left: whole bubbles or large fragments of their shells; note air cavities and escaping air bubbles. Right: almost entirely fragments of bubble-walls.

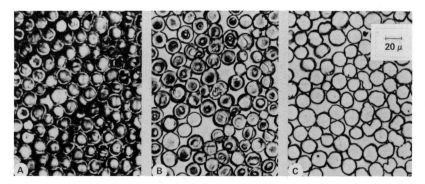

FIGURE 4-5. Experimental acrylic fibers, all from the same tow, but after different subsequent treatments, are shown as microtomed cross sections. (See also Figure 1-1.)

RECOGNITION OF PHASES

Under transmitted light, a piece of glass or a crystal is recognized with the naked eye as a single, homogeneous phase by its transparency and clarity. Heterogeneity caused by variation in refractive index due to insufficient mixing is manifested as schlieren* (transparent streaks or waves). Discontinuities such as boundaries between particles of the same or different phase are resolved as such, provided that the resolving power of the eye, the camera, or the microscope is sufficient for the purpose and the other attributes for visibility are adequate. If the boundaries are between contingent particles or pieces which are physically and chemically alike (let us say they are all particles of quartz, or all particles of silica glass), then there is still only a single phase. But if there are two or more different crystal structures present, even though they are of the same chemical composition (let us say particles of quartz *and* of tridymite or of cristobalite), then there are two or more different phases. So differences in crystalline structures (including polymeric ones) are detected by diffraction patterns of light, electrons, or x-rays, and also by optical and crystallographic properties[5] according to microscopical techniques.

STRUCTURE vs. MORPHOLOGY IN CRYSTALLOGRAPHY

By both x-ray and optical methods, crystallographic *structure* is classified by placement in either the isometric, tetragonal, hexagonal, orthorhombic, monoclinic, or triclinic crystal system, as indicated in the vertical columns of Figure 4-6.[6] The classification is based upon the relative spacing of the units (molecules, radicals, or ions) along the principal axes, and on the angles between the axes. "Form" in classical crystallography[7] definitely describes one kind of face, as specified by symbols such as Miller indices ("equant" column, Figure 4-6). "Form" has so many other meanings[8,9] that unless specified it could

---

*From the Greek, meaning a streak or layer; here, any fluctuation *within a phase* that changes the local density and so causes local refraction of light rays.

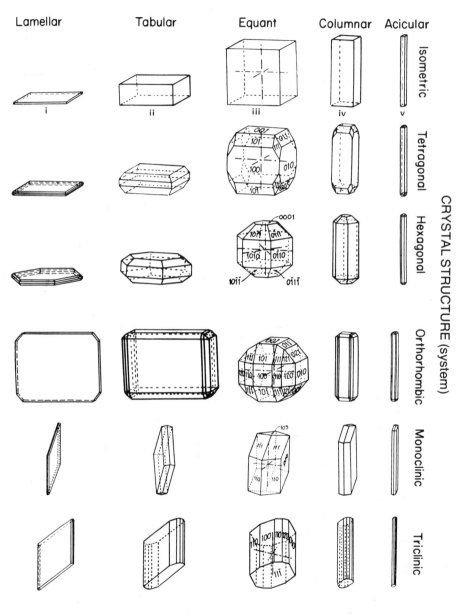

CRYSTAL MORPHOLOGY (habit)

FIGURE 4-6.   Crystal structure and morphology.

be ambiguous. Ambiguity is especially present when "form" is used generally for structure, morphology, or both. Indeed, for *each* of the six kinds of structure designated in Figure 4-6, there are five kinds of polygonal *morphology*—that is, habit. *Habit* is the shape of a polygonal crystal as determined by the kinds and relative sizes of faces present.[10]

Variations in habit in any of the crystallographic systems are equant, tabular, lamellar, columnar, and acicular (Figure 4-6). Common terms, such as flat and elongate, are acceptable, but flattened and elongated imply structural distortion, which in principle is not the case. That is, morphological variants (including habit) in any of the isometric tetragonal, hexagonal, orthorhombic, monoclinic, or triclinic structural systems *do not themselves* affect face angles, space lattice, x-ray pattern, optical properties, or melting point.

Figure 4-6, diagrams i-v, being in the isometric system, have equal spacings of units along the three mutually perpendicular axes represented in iii. The difference between iii (equant) and iv or v is that iv and v represent columnar and acicular crystals which contain *many more units* along one direction than along the other two. Likewise, i and ii represent equal (isometric) spacing between units, but they differ from iii, iv, and v in having obviously *fewer units* along one or two of the principal directions. Therefore the relative lengths of edges are not related to structure or to the unit cell, although the *angles* between edges are related to structure and symmetry (thus the value of goniometry in studying structure). Although habit does not affect intrinsic properties such as solubility, it *does* affect rate, such as rate of dissolution. Thus elongate or flat crystals dissolve more quickly than equant ones, simply because flat crystals offer more surface per unit weight than equant crystals.

Crystals which do not have the chance to develop faces fully are described as skeletal, dendritic, mossy, and so forth.[11] A hedrite[12] is a unit more fully developed than a dendrite, but less than a perfectly crystalline polyhedron. A spherulite (Figure 4-7) is an aggregate of unit, acicular crystals radiating from a common center. Crystalline grains are aggregates in which single crystals meet their neighbors in a haphazard way to produce *rough* polyhedral shapes. Their surfaces are *not* crystallographic faces; they are products of fortuitous packing.[13]

Size is another important feature of morphology. Among crystals,

FIGURE 4-7.   Spherulite of polyoxyethylene from the melt. Acicular crystals
radiate from the center. Also note the growth rings and shrinkage cavities.

the size usually is small when conditions favor rapid crystallization
(e.g., quick cooling) from a large number of nuclei. Conversely, large
sizes result from conditions favoring continued growth from a relatively small number of nuclei. Smaller crystals dissolve (and otherwise
react physically or chemically) faster than larger crystals, because they
present a greater specific surface to the reagent.

SEPARATION OF PHASES

In order to gain a full understanding of one phase among others,
separation may be necessary. It may be either tangible or intangible.
Tangible separation is represented by such macro- or micromethods as
hand picking, preferential solubility, fractional crystallization, selec-

tive gravitation, or centrifugation. Amenability to tangible separation depends not only on availability of a separatory method, but also on adequate supply of sample to yield sufficient amounts of separated phase for description, analysis, and test. The precautions to be taken center on avoiding change in physical phase as a result of such separatory methods. Care should be exercised to avoid the inadvertent addition of modifier, such as a solvent (which might act as a plasticizer).

Intangible methods, *in situ*, are represented by such methods as macroscopy, microscopy, diffraction, or spectrometry. The disadvantage of intangible separation is that it may be difficult to find a microtest of practical properties which may be performed *in situ*. There are, however, microtests for hardness,[14] refractive index,[15] and strength.[16] Amenability to intangible separation depends not only on having a method of distinguishing among the phases, but also on having a method of determining differential properties. A case in point is given by two commercial batches of poly(methyl methacrylate) intended for optical exploitation. Both were satisfactory when used as separate batches, but when mixed, the molded products manifested distinct schlieren patterns. (Such streaks are also seen in inorganic glass sometimes, and are known to be caused by incomplete mixing of two glasses which differ even slightly in refractive index.) Thus the observation of schlieren was both a method of distinguishing between two phases and of detecting a difference in properties. But the manufacturer insisted that the two batches had "the same" refractive index because the indices agreed to the fourth decimal place. By employing a more sensitive refractometer, a difference was found in the fifth decimal place. This is more than enough difference to cause all the trouble. The manufacturer still insisted that he had distilled "all" of the xylene from the batch of poly(methyl methacrylate) made by polymerization in xylene, but the occurrence of schlieren patterns proved the presence of xylene—by intangible separation of phases.

SUMMARY

Ice and water are two different phases of the single component H_2O and have entirely different properties by which we recognize them. Similarly, crystalline polystyrene and amorphous polystyrene are two different phases of the same substance, with entirely different properties. Obviously the resinographer has to distinguish between different phases of a single component, and to appreciate the effect that different proportions of those phases will have on the properties of the polymer.

If salt is dissolved in water, we have only one phase, but two components. Here we distinguish the single phase from its constituents by considering it as a system of components, and we can explain the properties of the solution on the basis of those components. Similarly, if we have two polymerizable components (which we shall call monomer A and monomer B) which form a solution, and this solution is then polymerized, the copolymer is a single phase, but we can distinguish it from its pure constituents by considering it as a system of components, and we can explain its properties on the basis of an interpolymer of those components. In a little different way, a liquid unimeric plasticizer A may be dissolved in a pure polymer B to have a single phase which has properties different from those of the pure components.

If two solid substances, A and B, are only partially soluble in each other, then we may have a solution of A in B, a solution of B in A, or a mixture or dispersion of the two saturated solutions, depending on the proportions of A and B. Obviously the resinographer has to recognize the two solution phases, and relate the overall properties of the material to the proportions and properties of those phases. Similarly, the resinographer has to find and identify all the phases in a polyphase plastic material in order to arrive at a complete description. There are structural aspects of each phase (amorphous, crystalline, anisotropic, colored, and opaque) and morphological aspects (size, shape, the distribution of these, and their changes under treatment) by which they can be recognized.

These matters are illustrated in this chapter by considering the rosin–rubber system, the poly(methyl methacrylate)–poly(ethyl acrylate) copolymer system, the poly(vinyl chloride)–tricresyl phosphate system, and several polyphase systems involving mineral fillers. Since the recognition of phases by means of their crystallinity (or lack of it) is new to many readers, a section on the methods of recognizing phases is included, covering crystallinity, the crystal systems, crystal habit, and anisotropy. The *separation* of phases, whether it be actual (tangible) separation to "stand apart and be counted" or just differentiation and estimation in the mind (intangible separation), is also an important consideration in this chapter. The general message of the chapter is that the study of *phases*, embracing their recognition, identification, and quantitative estimation, is at least as important to resinography as the investigation of chemical composition and molecular structure on Level I.

SUGGESTIONS FOR FURTHER READING

ACS symposium on morphology of polymers (T. G. Rochow, ed.) at Los Angeles, Calif., April 4–5, 1963, in *J. Polymer Sci., Part C*, No. 3 (1963); also as a monograph by the Interscience Div., John Wiley and Sons, Inc., New York, N.Y. 10016 (1963); of 16 papers, especially the following six on the crystalline phase: C. S. Hsia Chen and D. G. Grabar, Morphological aspects of polymerization in the solid state, pp. 105–107; N. K. J. Symons, The morphology of crystalline polymers, with especial reference to single crystals grown from the molten state, pp. 109–116; F. P. Price, Kinetics of spherulite formation, pp. 117–119; H. D. Keith, Mechanisms and kinetics of spherulitic crystallization in high polymers, pp. 121–122 (abstract only); F. R. Anderson, Fracture studies of isothermally bulk-crystallized, linear polyethylene, pp. 123–134; K. Sasaguri and R. S. Stein, Relationship between morphology and deformation mechanisms of polyolefins, p. 135 (abstract only).

REFERENCES

1. C. W. Hock and A. N. Abbott, Topography of pressure-sensitive adhesive films, *Rubber Age* **82**, 471–475 (Dec. 1957).
2. W. D. Stonecipher and R. W. Turner, Rosin and rosin derivatives, in *Encyclopedia of Polymer Science* (H. F. Mark, ed.), Vol. 12, pp. 139–161, John Wiley and Sons, Inc., New York, N.Y. 10016 (1970).

3. *Kirk-Othmer Encyclopedia of Chemical Technology* (A. Standen, ed.), 2nd ed., Vol. 17, pp. 379–391, Interscience Div., John Wiley and Sons, Inc., New York, N.Y. 10016 (1968).
4. T. G. Rochow and E. G. Rochow, companion monograph on microscopy, in preparation.
5. A. N. Winchell and H. Winchell, *The Microscopical Characters of Artificial Inorganic Solid Substances: Optical Properties of Artificial Minerals, Academic Press*, New York, N.Y. 10003 (1964).
6. A. N. Winchell, *The Microscopic Characters of Artificial Inorganic Solid Substances or Artificial Minerals*, John Wiley and Sons, Inc., New York, N.Y. 10016 (1931).
7. E. E. Wahlstrom, *Optical Crystallography*, 4th ed., p. 5, John Wiley and Sons, Inc., New York, N.Y. 10016 (1969).
8. E. M. Chamot and C. W. Mason, *Handbook of Chemical Microscopy*, 3rd ed., Vol. 1, pp. 173, 336, 367, and 410, John Wiley and Sons, Inc., New York, N.Y. 10016 (1958).
9. E. M. Slayter, *Optical Methods in Biology*, pp. 109, 336–338, Interscience Div., John Wiley and Sons, Inc., New York, N.Y. 10016 (1970).
10. F. C. Phillips, *An Introduction to Crystallography*, 4th ed., pp. 11–12, John Wiley and Sons, Inc., New York, N.Y. 10016 (1971).
11. *AGI Glossary of Geology and Related Sciences*, 2nd ed., American Geological Institute, Washington, D.C. 20037 (1960).
12. T. G. Rochow, article on resinography, in *Encyclopedia of Polymer Science* (H. F. Mark, ed.), Vol. 12, John Wiley and Sons, Inc., New York, N.Y. 10016 (1970).
13. Grain boundary, *Glossary of ASTM Definitions*, 2nd ed., p. 209 American Society for Testing and Materials, Philadelphia, Pa. 19103 (1973).
14. ASTM designation E384, *Standard Method of Test for Microhardness of Materials*, ASTM Standards, American Society for Testing and Materials, Philadelphia, Pa. 19103 (annual).
15. ASTM designation D542, *Index of Refraction of Transparent Organic Plastics*, ASTM Standards, American Society for Testing and Materials, Philadelphia, Pa. 19103 (annual).
16. ASTM designations D1445, *Test for Breaking Strength and Elongation of Cotton Fibers (Flat Bundle Method)*; D1294, *Test for Breaking Strength of Wool Fiber Bundles, 1-in. Gage Length;* D2101, *Test for Man-Made Fibers Taken from Filament Yarns and Tows;* also ASTM designation D638 *Test for Tensile Properties of Plastics,* ASTM Standards, American Society for Testing and Materials, Philadelphia, Pa. 19103 (annual).

5

LEVEL III: SURFACES

We turn now from molecules and phases to the domain of *surfaces* — the domain of interfacial phenomena and of colloid chemistry. When we considered molecules on Level I, we thought of them as being exposed to similar molecules on all sides; when we considered phases on Level II, we thought of them as surrounded by more of the same phase or by associated phases. But now we come to the situation where some of the molecules which comprise the material are exposed to *different* molecules — alien molecules of gas, or liquid, or even solid. The boundary is traditionally called a *surface* if it separates a condensed phase from the air, and an *interface* if it separates the condensed phase from a liquid or a solid, but the same principles are involved in both situations. As before, we must consider the surface or interface not only in terms of structure and composition, but also in terms of morphology, condition, and behavior.

Theoretically the surface or interface is only one molecule thick. However, if there is any interaction or polarization, a layer more than one molecule thick may form. It is essential to the reasoning (and to the deductions made from it) that the boundary layer not grow thick enough to be detected as another phase. The principal idea for Level III is that the external molecules of the material are in contact with the same kind of molecules on one side and with utterly different molecules on the opposite side. Therefore they are under more than one kind of molecular influence.

Our first big question about surfaces is: What kind of *structural units* are there? Is one dealing with discrete molecules, domains, parti-

85

cles, flocs, fibers, films, sheets, or even bulkier articles? Are the structural aspects any different as we progress from solutions to suspensions, foams, emulsions, lattices, polyblends, laminates, composites, and filled plastics?

The second set of questions is about the morphological aspects: What sizes and shapes are we dealing with? What is their distribution? Is topography important? What is the specific surface (or interface) per unit weight? Is the composition at the surface different from that of the interior? If so, how does the difference affect the properties? What is the condition after exposure to air, to the weather, to intense sunlight or heat or radioactive radiation? Does the resulting condition affect the behavior with changing time, or wear, or mechanical fatigue? Some of the possible answers to many of these questions are outlined in Table 5-1.

Rosin again serves as a first example. Rosin "sizes" are used on paper; their presence makes the difference between writing paper and blotting paper. But the properties and behavior of rosin that make this difference are due to more than just the different molecules making up the rosin, or the bulk phase considered in the last chapter. The difference lies also in the architectural units of the rosin size itself. Those units are small, separate particles, and they are converted to small agglomerates (called *flocs*) by adding papermaker's alum, $Al_2(SO_4)_3$. The flocs are substantive in the sense of a dye; that is, practically all of the rosin-bearing flocs seek out and stick to the paper fibers, in spite of the low proportion of rosin to fibers and the low concentration of fibers suspended in the water contained in the papermaker's beater. After the mat of paper fibers is formed, and during the final calendering process, the rosin-bearing particles are squeezed flat and glazed over the paper fibers and into crevices. Thus the size offers tremendous *specific surface* to ink, so as to keep it from feathering into and between paper fibers. The composition of the particles includes some aluminum and sodium as rosinate soaps, the proportions depending on the proportion of alum added to the commercial rosin size. While some of the characteristic properties and behavior of rosin sizing may be attributed to the presence in part of these soaps on Level I, the principal characteristics must be attributed to the particulate structure, the flocculent morphology, and the great specific surface of the whole rosin complex as treated on Level III. Figure 5-1 helps describe the rosin-sizing phenomenon. It also indicates how the flocs may be used as carriers of other paper-finishing agents, such as waxes.

Man-made resins such as melamine-formaldehyde polymer may also be given colloidal and substantive properties so as to collect on

FIGURE 5-1. Electron micrograph showing flocs of experimental uranyl rosinate adhering to fribrils of air-dried pulp.

TABLE 5-1.

Resin, etc.	Structure (kind)	Morphology (extent)
Natural, recent	Particles; films	Specific surface
Natural, modified	Films	Specific surface
Natural, fossils	Bulk	Pieces, carvings
Modified fossils	Solute; films	Classical painting
Natural polymers	Fibers; pulp	Cloth, paper, etc.
Regenerated polymers	Fibers, foil, sponge	Cloth, film
Modified polymers	Fibers, foil, moldment	1, 2, 3 dimensions
Synthetic condensates	Particles, moldment	Shapes, sizes, pores
Addition polymerization	Fibers, film, particles	Sizes, shapes
Rubber, elastomer	Tubes, moldment, sponge	Sizes, shapes, pores
Gums	Film; adhesive	Thickness
Starches	Granules, film	Shape; size; hilum
Starches, modified	Film, adhesive	Degree modified

Typical Materials in Terms of Their Constituent Surfaces (Level III)

Composition	Condition	Properties	Behavior
Variable	Dispersion and coagulation	Sizing; adhesive	Change with environment
Soaps, esters	Dissolution in H_2O; soluble	Dispersive	Useful change
Still obscure	Original, aged	Insoluble in H_2O, solvents	Stable
Still obscure	From solution in oils	As applied	Classical
Vegetable, animal, mineral	As separated, cleaned	Natural	Natural
Cellulose, protein	As spun, cast, etc.	Of high specific surface	Stable?
Cellulose esters; others	As spun, cast, etc.	Of high specific surface	Stable?
Continuous resin	As cured	Insoluble	Stable; weatherable
Variable	As extruded, cast, precipitated	Frictional, etc.	With time, etc.
Variable	As extruded, cast, precipitated	Rubbery	In weather, time
Natural or modified	Spread solution	Adhesive with H_2O	Not H_2O-proof
Sheaves of molecules	Levigation	Typical pastes in H_2O	Cookable
Granules	Levigation, acid or ferment	More soluble in H_2O	Cooked

paper-making fibers, even though they are suspended in large proportions of water. In Figure 5-2, an electron micrograph, particles of melamine-formaldehyde resin are shown in the aqueous colloidal state. The particles vary in diameter from about 0.06 μm to 0.005 and less, with the average about 0.01 μm.[1] The resin is cured by heat during the stages of drying the paper. The very widely distributed resin gives the paper superior fold resistance[2] and exceptional strength while wet. Similarly, urea-formaldehyde resins form a substantive colloid in acidified water. Or a sulfonic group may be attached to urea which, with formaldehyde, forms an anionic colloid which flocs with alum. Both types of flocculated urea-formaldehyde resins are used to give strength to wet paper.[3]

Besides suspensions, colloidal systems include emulsions, latices, and polyblends (Figure 5-3). In all of these the composition,

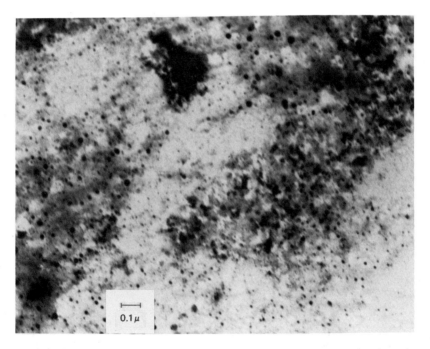

FIGURE 5-2. Dry residue of aqueous colloid of acidified melamine-formaldehyde resin, directly illuminated by transmitted electrons. Electron microscope (33,000 ×).

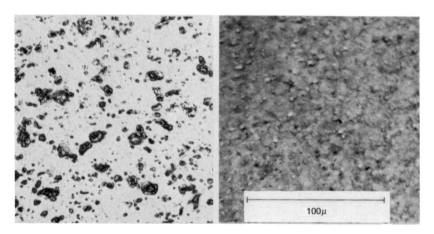

FIGURE 5-3. Two "alloys," high in polystyrene containing small particles of elastomer, illustrating also two methods of preparation and two methods of lighting. Left: polished and etched (benzene in methyl alcohol ca. 1:4); bright-field reflected illumination. Right: microtomed thin section; bright-field, slightly oblique transmitted illumination.

properties, and description of the interfaces are to be classified separately on Level III from the resinography of the bulk polymers in separate phases on Level II. In some instances of emulsions[4], however, the so-called "interface" may actually be *another phase*[5] composed of emulsifying agent plus the bulk components. As a corollary, some polyblends may have zones intermediate in composition between those of the two principal phases. In some fluid suspensions, as in paint, the particles of pigment may have been treated with a very thin layer of a surface-active agent that aids in producing suspension, thereby prolonging shelf life or improving service. The description of such surfaces is difficult, but is nevertheless desirable in understanding what goes on.

Ionic resins may be produced in the form of granular spheres or beads for use in ion-exchange processes to purify water, beer, sugar syrups, pharmaceuticals, and chemicals, or to recover valuable materials from waste. While such resins depend fundamentally on their chemical activity, they also depend on their specific surface for their efficiency and long life.

Specific surface is also of key importance in the behavior of permeable or semipermeable membranes. Within sponges and cellular plastics, the thin walls should be stable to the gas involved in the expanding process, and to the practical environment. The walls must also be permeable to water and solvents, and of course the walls should be wettable. In all cases the shapes, sizes, and distribution of the structural cells are important[6] (see Figure 5-4).

The significant units involved on Level III vary in size from molecules, colloidal particles, granules, or pellets to large pieces. Articles such as a solid ball may have only one surface exposed to the environment. But a hose, tube, or pipe has one surface exposed to the atmosphere and the other to the fluid it carries, such as gasoline. Other containers may also carry a particulate solid which may be abrasive or corrosive. Films, foils, and sheets also have two principal surfaces; one surface may be exposed to a different environment than the other. For example, the two sides of the polyester foil shown in Figure 5-5[7] are different; one was cast against a smooth solid and the other hardened in rougher form since it was exposed to air when cast. In this instance the two surfaces differ in texture so much that, when the foil is folded one way, it may be felt to slip more easily than when folded the other way.[8] The effect may be more pronounced if the polymer has crystallized, as illustrated in Figure 5-6. The spherulitic crystalline habit also is involved, because it manifests in three dimensions the spiny characteristics of acicular crystallites in spherulites. The surface cast against a smooth flat substratum shows an essentially smooth, two-dimensional surface, even though it has crystallized (Figure 5-6).

The actual crystallization of a polymer such as nylon depends not only on the crystallizability of the polymeric molecules involved, but also on the influence or epitaxy[9] of the surface of the mold. The conditions of crystallization, such as temperature, time, seeding, and agitation or shock, also are involved. They affect the number and size of nuclei and the rate of growth. Such conditions can be quite different for interfacial molecules than for interior ones, especially when casting thick specimens against very hot or very cold walls.

The surface of a thermoplastic material which is subjected to enough heat, solvent, or pressure will take on the texture of the contacted solid. Such indeed is the purpose of replicating techniques in light microscopy and in electron microscopy. However, replication

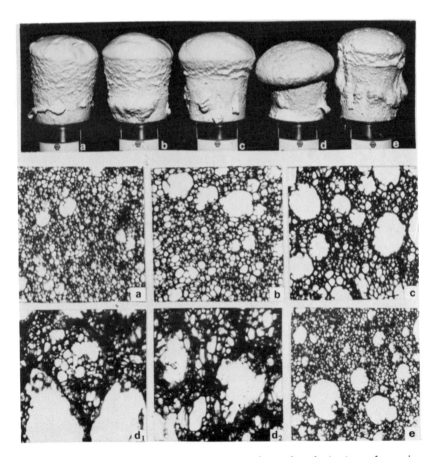

FIGURE 5-4.   Effect of increasing concentrations of surfactants on foam rise. The samples of foam are shown in the top photo, while the bottom photos show the typical skeletal structures (18×) corresponding to these. The amount of auxiliary surfactant increases from left to right, with no surfactant present in (a). Note that (b), (c), and (d) show increasing drooping and roughness of skin, while the (e) again shows a firm rise. The drooping is associated with coalescence of cells and partial collapse due to loss of blowing agent.

Frames d_1 and d_2 are from the coarse outer layer and the finer inner layer, respectively, of foam d. The foam pictured in frame d_1 consists of large voids surrounded by a continuum of resin in which the remnants of smaller cells are embedded. Note the fine cell structure of foam e.[6]

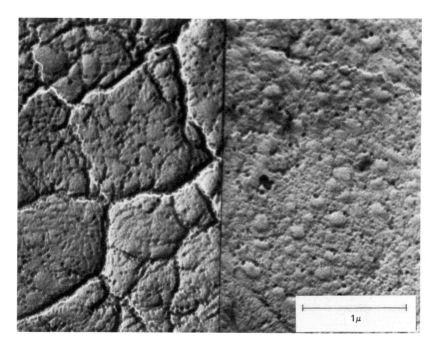

FIGURE 5-5. Commercial film of polyester showing a difference in the texture between the two sides. Electron micrograph of a positive silica replica; no shadowing metal. Right: smoother side, cast against a smooth solid. Left: rougher side, cast against air.

may be a nuisance, as when the plastic shield of a portable television set happened to be pressed against the upholstery of a hot automobile. Here there was accidental replication of the grain of the polyvinyl textile due to the migration of a plasticizer that had collected at the surface of the shield and had softened the surface.

Moving solids present a challenge to the surfaces of plastic bearings, such as those of nylon, where the surface molecules must act as lubricant, while the interior molecules must provide strength to the bearing.

Surfaces may be specially cast, embossed, or otherwise manufactured for specific purposes. As illustrated in Figure 5-7, a transparent sheet may be made so that the surface exposed to incident light may scatter the light softly so as to reduce, if not eliminate, reflected images. This represents an improvement in transparent covers for pictures or other objects to be viewed through the cover. The surface shown in Figure 5-7 is covered with tiny, curved facets. If the facets are made large enough, the surface may be rendered retroreflective for use in projection screens, traffic signals, danger signs, and decorative effects. Retroreflective surfaces may be made by embossing lenses or prisms, or by embedding transparent spheres. Various other optical or decorative effects may be made on surfaces or interfaces during casting, extruding, or molding.

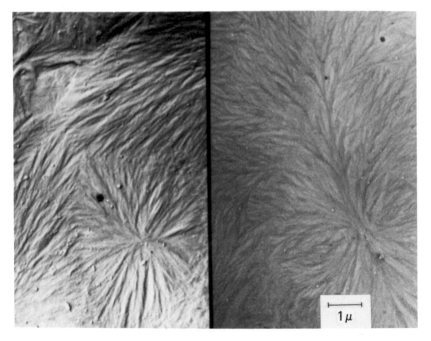

FIGURE 5-6. Commercial film of nylon showing differences in texture between the two sides. Electron micrographs of positive silica replica; no shadowing metal. Right: smoother side, cast against a smooth solid. Left: rougher side, cast in air.

FIGURE 5-7. Surface of nonreflective glass, prepared for photography by coating surface with vaporized aluminum. Photomicrographed with slightly oblique reflected light.

Films such as these of varnish, paint, or lacquer have airsurfaces which may be derived not only from the nature of the material, but also from the method of application. Examples of the latter are brush marks (Figure 5-8), wavelets, and "orange peel." The airsurface encounters the weather or other environment, whereas the undersurface of a protective film must adhere and stay adhered to the substratum.

Permanently adhesive films which are used between the layers of laminates such as plywood must have surfaces which stick to the plies. Similarly, boards composed of wood chips or wood pulp and plastic, as well as sheets made of fabrics, textiles, paper, or foil, depend on adhesive surfaces to keep them together. Conversely, some temporary films such as those used to protect adhesive surfaces until the object is applied in a permanent position must possess a surface which will easily peel away. A corollary requisite is that the temporary recipient surface must reciprocate in the peeling process, and that the permanent recipient surface must cooperate in the adhesive stage. Some adhesives require high or long pressure (as with hardenable resins), whereas others are so-called pressure-sensitive. Whatever the prescribed conditions, the bonding interfaces must match and wet each other, and these separate requirements must be met by the same or different interfacial phases.[10] At the same time, the interior molecules must adhere to the interfacial molecules, and in both positions the molecules must adhere to each other for the intended purpose.

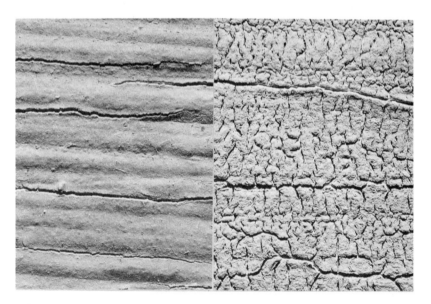

FIGURE 5-8. Painted wood panels showing cracking in brushed oil-paint film. Oblique, reflected light was used (approximately 11×). Left: straight cracks, following brush marks and grain of wood. Right: straight cracks as at left, but with checking.

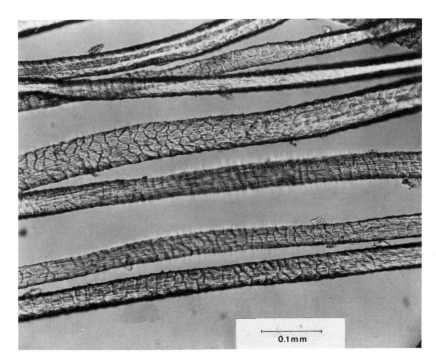

FIGURE 5-9.   Surgical grade wool fibers, focused to show the surfacial scales so functional in spinning and in felting. Mounted in glycerine-water solution ($n = 1.371$). Transmitted light.

The surfaces of fibers and filaments are also important. Such surfaces may be natural, as in wool (Figure 5-9), or the surface may be modified, as in mercerized cotton, or it may be completely man-made, as in viscose rayon. The topography is characterized microscopically by longitudinal and cross-sectional views. The surfacial layer may sometimes be peeled away[11] from the core, as illustrated in Figure 5-10. Sometimes the layer is thick enough to be seen in cross section, and is called the skin (as opposed to core), as in Figure 5-11.

Extruded rods, as well as extruded fibers, carry surfaces which are characteristic of the extruding and drawing process. Such rods are often broken into pellets for molding. The resulting moldment bears a surface which is a negative replica of the surface of the mold. But after

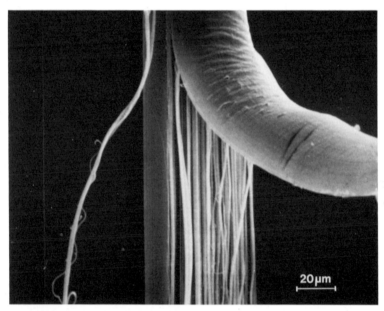

FIGURE 5-10.   Fibrillar structures of a highly drawn polypropylene filament. Peeling and scanning electron micrography by P. Tucker.

FIGURE 5-11.   Skin-core structure of an experimental acrylic fiber; an electron micrograph of an ultra-thin section by M. C. Botty.

annealing above the glass transition temperature, the compression moldment may show a peculiar reticulation reminiscent of the sizes and shapes of the original extruded pellets (Figure 5-12). Evidently the original pellets "remember" their identity by means of their individual surfaces. In such cases, at least, there is evidence that the original surfacial molecules are different from the internal ones, and are coherent enough to draw away from their neighbors and try to resume their previous positions in the "skin" of the pellet.

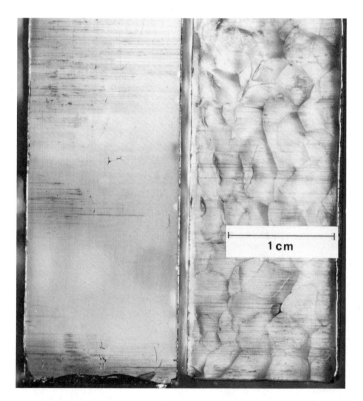

FIGURE 5-12. Photograph of a commercial poly(methyl methacrylate) (left) after experimental compression molding and (right) after annealing at 116°C. The reticulation is definitely related to the size and shape of the original molding grain.

SUMMARY

Regardless of the actual atomic or molecular composition of a material, and quite apart from the actual phases present in it, the *surface* of the material sometimes becomes the all-important consideration. This is particularly true of paints, of sizes for paper and plaster, of permeable and semipermeable membranes, and of adhesives. In fact, the surface of any liquid or solid must be considered as something special, because the molecules which comprise the surface are all attracted inward by like molecules, but they face molecules of a different sort (friendly, unfriendly, or even hostile) outside the surface.

While we think of only one layer of molecules of a substance at a surface or interface, the special forces which exist there will usually polarize or orient adjacent molecules as well. As a result, we actually are dealing with a "skin" which may be several or many molecules thick, and which grades off imperceptibly into the interior of the phase. The special forces within this skin may draw it together by surface tension, so that it presents minimum area. A forcibly flattened out particle or pellet of resin may "remember" its previous conformation because of those still-existing special associations, so that when given a chance (by warming, or through relaxation produced by a solvent), the particle tries to resume its original conformation.

The science of surfaces (colloid chemistry) is very different from analytical or synthetical chemistry, for it deals with wettability, adhesion, abhesion, and all the other surface phenomena rather than with composition or structure. These special phenomena are illustrated very well by a consideration of paper sizing, where rosin or man-made resins are prepared in a special morphology and applied in a special way to cellulose fibers during the paper-making process. When properly introduced and taken up by the fibers, and then spread out and glazed by hot rolling, the sizing material confers special properties to the paper: ink stays on the surface instead of spreading to the interior, the surface stays smooth and glazed, and the paper is much stronger. Special sizing materials also confer wet strength and fold-cracking resistance. All the benefits come from a very small proportion of sizing material, because only the surface counts.

Fibers and very thin sheets (foils) are also highly dependent on surface characteristics for their utility. This is illustrated in this chapter by photomicrographs of wool and of specially processed man-made fibers. The extrusion of man-made fibers and the molding of plastic objects introduce further considerations of surface, for the molds and dies which are used exert their influence on surface orientation and crystallization, and (at the very least) leave marks which have their effects later.

The area where surface properties take precedence over all others to the greatest extent is that of adhesives. Here the mutal surface interaction of structural solids with liquid adhesives involves many considerations of wettability, strength of the adhesive bond after application and processing versus strength of the materials themselves, resistance of the bond to water, weathering, and solvents, changes due to aging, and so on. These matters are of supreme importance in the making of plywood and laminated plastics, the reinforcing of resins by glass fibers and powdered fillers, the strengthening and toughening of rubber by carbon black, and the protection of anything by paint or varnish. The products and the processes that produce them can be monitored by the resinographic techniques recounted in each instance.

SUGGESTIONS FOR FURTHER READING

Resinography of Cellular Plastics (R. E. Wright, ed.), ASTM Special Technical Publication 414, A symposium of seven papers on foams, i.e., dispersions of gas in a solid resin or elastomer, American Society for Testing and Materials, Philadelphia, Pa. 19103 (1967).

ACS Symposium on morphology of polymers at Los Angeles, California, April 4–5, 1963 (T. G. Rochow, ed.) *J. Polymer Sci., Part C*, No. 3, 1963; reprinted by Interscience Div., John Wiley and Sons, Inc., New York, N.Y. 10016 (1963). Of special interest are the papers by E. B. Bradford and J. W. Vanderhoff, The morphology of synthetic latexes, pp. 41–64; E. H. Erath and Morton Robinson, Colloidal particles in the thermosetting resins, pp. 65–76; and J. P. Berry, The morphology of fracture polymer surfaces, pp. 91–101.

REFERENCES

1. T. G. Rochow, The microscopy of resins and their plastics, *ASTM* Symposium on Light Microscopy, ASTM Special Technical Publication 143, American Society for Testing and Materials, Philadelphia, Pa. 19103 (1952).
2. I. Bursztyn, The wet strength of paper, *British Plastics*, pp. 299–304 (July, 1948).
3. H. P. Wohnsiedler, Urea-formaldehyde and melamine-formaldehyde condensations mechanisms, *Industrial and Engineering Chem.* **44**, 2679–2686 (1952).
4. T. G. Rochow and C. W. Mason, Breaking emulsions by freezing, *Industrial and Engineering Chem.* **28**, 1296–1300 (1936).
5. T. G. Rochow, article on resinography, in *Encyclopedia of Polymer Science and Technology* (H. F. Mark, ed.), Vol. 12, Interscience Div., John Wiley and Sons, Inc., New York, N.Y. 10016 (1968).
6. *ASTM Symposium on Cellular Plastics* (R. E. Wright, ed.), ASTM Special Technical Publication 414, American Society for Testing and Materials, Philadelphia, Pa. 19103 (1967).
7. T. G. Rochow, Resinography of high polymers, *Analytical Chem.* **33**, 1810–1816 (1961).
8. F. Eirich and H. F. Mark, *Congrès International de Microscopie Electronique*, Paris, Sept. 14–22, 1950.
9. P. Tucker, School of Textiles, N.C. State University at Raleigh 27607, unpublished lecture at University of Leeds, England, March 14, (1974).
10. C. W. Hock, The morphology of pressure-sensitive adhesive films, *J. Polymer Sci. Part C* **3**, 139–149 (1963); also published in the monograph on *Morphology of Polymers* (T. G. Rochow, ed.), ACS Symposium, Los Angeles, Apr. 4–5, 1963, Interscience Div., John Wiley and Sons, Inc., New York, N.Y. 10016 (1963).
11. R. G. Scott, The structure of synthetic fibers, *ASTM Symposium on Microscopy* (F. F. Morehead and R. Loveland, eds.), ASTM Special Technical Publication 257, American Society for Testing and Materials, Philadelphia, Pa. 19103 (1959).

6

LEVEL IV: MATERIALS

We come now to the ultimate product, be it fiber, plastic, rubber, paint, or whatever, and we wish to consider the product on the level of the whole material, no matter how many phases or how many kinds of molecules it may contain. This is the level of industrial design, of materials engineering, and of practical testing. It is of interest to engineers, technologists, industrial and commercial people, and of course the consumer.

To engineers and other technical people, the most important practical properties of a material are those of the bulk material as a whole, properties which can be guaranteed by test. Accordingly, such tests have been established on a nationwide scale, so that *producers and consumers together* can come to a common understanding on what the material is, what it can do, and how it will be affected by conditions encountered during use. For this purpose the American Society for Testing and Materials is the definitive body, as indicated in previous chapters.

A particular ASTM test may measure a single property, or several at once. A few well-known types of tests are tensile strength (D638), flexural strength (D790), impact strength (D256), Rockwell hardness (D785), compressive properties (D695), deflection temperature (D648), flammability (D635), water absorption (D570), specific gravity and density (D792), electrical resistance (D257-61), and refractive index (D542).[1]

Some of the questions to be answered on Level IV are: (1) What is the *kind* of structure, according to the plan set forth in Table 6-1?

TABLE 6-1.

Resin, *etc.*	Structure (kind)	Morphology (extent)
Natural, recent	Varnish, paint, etc.	Proportions, distribution
Natural, modified	Varnish, paint, etc.	Proportion, distribution
Natural, fossils	As carved, etc.	Proportion, distribution
Modified fossils	Varnish, paint	Proportion, distribution
Natural polymers	Fibers, fabric, etc.	Proportion, distribution
Regenerated polymers	Fibers, fabric, etc.	Proportion, distribution
Modified polymers	Fibers, fabric, etc.	Proportion, distribution
Synthetic condensates	Composites	Moldments
Addition polymers	Alone, polyblends	Moldments
Rubbers, natural, synthetic	Alone, polyblends	Moldments
Gums	Foods, adhesives	As manufactured
Starches	Foods, adhesives	As manufactured
Starches, modified	Foods, adhesives	As manufactured

Typical Materials in Terms of Themselves (Level IV)

Composition	Condition	Properties	Behavior
Variable	As mixed, milled	Overall	With time, use
As modified	As mixed, milled	Overall	With time, use
Natural, composite	Things such as jewelry	Overall	With time, use
As modified	As mixed	Overall	With time, use
Cotton, wood, pulp	Suspension, impregnation	Strong (tires, etc.)	Standard
Viscose, cellophane	Suspension, laminates	Strong (tires, etc.)	Sometimes better
Acetate, rayon, etc.	Mixed in fabrics	Lustrous, smooth	Thermoplastic
As marked, published	Compression molded	ASTM standards	ASTM standards
Polystyrene, poly(methyl methacrylate), polyethylene, polypropylene	Cast, injected, extruded	ASTM standards	ASTM standards
Plus accelerator	Rolled, extruded, etc.	ASTM standards	ASTM standards
As manufactured	As sold, used	ASTM standards	ASTM standards
As manufactured	As sold, used	ASTM standards	ASTM standards
As manufactured	As sold, used	ASTM standards	ASTM standards

(2) What is the composition in terms of phases? (3) What are the *qualitative* overall properties (stiff, flexible, or extensible; transparent, translucent or opaque; thermoplastic or thermoset, etc.)? (4) What are the *quantitative* overall properties (tensile, impact, and flexural strength; wear resistance; heat resistance, etc.)?

In terms of morphology, the questions are: (1) What are the dimensions and what is the uniformity of structure (e.g., diameter or denier of fiber; thickness of a sheet; dimensions of a repetitive article)? (2) What is the dimensional stability in terms of the test treatment or use? (3) What use tests are to be made (e.g., natural exposure vs. weatherometer; wear vs. abrasion tests; use vs. flexing or other tests; variable thermal or atmospheric use vs. tests in an oven, refrigerator, or humidor, etc.)? (4) What is the behavior over periods of time under the aforesaid conditions?

MULTIPHASE PLASTIC MATERIALS

Sometimes a plastic must be clear, transparent, and isotropic in order to accomplish its intended purpose, as when poly(methyl methacrylate) is used for molded optical lenses. Far more frequently, a plastic composition must have mechanical strength and endure impact and abuse, and in such cases a *polyphase* composition is employed in order to achieve superior mechanical properties. The principle involved is a familiar one: a glass rod, composed of a single isotropic phase, is easily broken after scratching because the surface scratch is propagated as a crack due to the intense concentration of stress at the apex of the crack. If particles of other solid phases are dispersed in the glass, however, the incipient crack soon encounters one of these solid particles, where the apex is blunted as the stress is distributed over a mass of material. So it is that porcelain, with its many dispersed phases of aluminates in a glassy matrix, is much more rugged and durable than glass. Similarly, dispersed particles of iron carbide greatly strengthen iron (by preventing slippage of the crystal planes of pure iron) and make steel out of it, and dispersed particles of silicon and various silicides strengthen aluminum (by precipitation hardening) and make aircraft alloys or automobile engine blocks out of it.

In fact, the workday world is filled with examples of polyphase materials that are stronger, better, more flexible, or more durable just because they incorporate several "specialist" phases into a synergistic composite: wood, with its strong fibers of cellulose embedded in resinous lignin; bone, with its particles of aragonite and apatite embedded in a cartilaginous matrix; "fiber glass" boats and fishing rods, in which strong fibers of glass are enclosed and protected from scratching by a matrix of organic polymer; rubber tires with their reinforcing particles of carbon black; paints with their pigments; and so on through countless examples within the realms of metallurgy, biology, and ceramics. In the realm of plastics and elastomers, the engineering materials derived from them are almost always hardened, strengthened, or rendered more durable by incorporating fibers, fillers, stiffeners, and reinforcing agents. Often pigments are milled into the matrix for color, and particles or flakes (of aluminum or bronze, for example) are added to shade the polymer from sunlight. Sometimes fillers are added to increase the density, or gas bubbles to decrease the density. The end result of all this is a *composite* material, and it is the job of the resinographer to identify the various phases in the material and to reconstruct the rationale that led to their incorporation. The specific illustrations and comments which follow will be of help.

If we think first of *films* based on resinous or polymeric material, we see that these could seldom do their jobs if they consisted of a single phase. Lacquer or varnish is spread as a surface coat on wood, paper, cloth, leather, or metal, giving a composite of improved performance. Paints have pigments, fillers, and sunshade components in a varnish vehicle, as explained. Even cellophane is coated or laminated to decrease its permeability to water vapor. Laminated materials such as plywood (especially prefinished decorative plywood) and kitchen-counter laminate provide other excellent examples of cooperative phases. The artificial leather of Figure 2-3 (Chapter 2) shows several layers of unwoven fabric, porous plastic, and pigmented polymer which make up the leatherlike structure. Figure 6-1 shows two samples of laminated paper in which the different layers have different functions.

Woven fabrics such as cotton duck or canvas have long been incorporated into thermosetting resins such as the phenol-formaldehyde type to give strength and resilience. The quiet timing gears of au-

THOUSANDTHS OF AN INCH AT 250 DIAMETERS

FIGURE 6-1. Photomicrograph of a polished cross section of several functional layers of resin-impregnated paper, taken by vertically reflected light.

tomobile engines provide a good example of the end use of such a composite material. For added strength or rigidity the fabric may be of inorganic glass fibers, either as a roving yarn or as a woven fabric, as in laying up boat hulls.

Fibers themselves may be of polyphase structure in order to accomplish particular purposes. Spun or drawn fibers of a single material often show a skin-plus-core structure, whether purposeful or not, as shown in Figure 6-2. Lately, some fibers are being made with one or more air canals to improve strength and increase bulk (Figure 6-3). Some dual-phase fibers have been introduced in order to obtain the crimp, curl, or shrinkage necessary for bulking (See Figure 6-4). Glass fibers frequently are coated to obtain adhesion, dyeability, or some other property. Nylon filaments for paint brushes nowadays are tapered to give the brush more of the properties of animal hair or bristles. Such man-made bristles may also be coated to improve wettability or

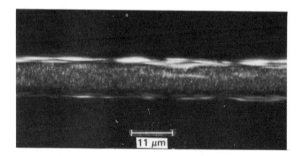

FIGURE 6-2. Experimental acrylic fiber. Photomicrograph by dark-field transmitted light, picturing an acrylic fiber as a highly oriented (strong) casing filled with more loosely packed (more dyeable) internal particles.

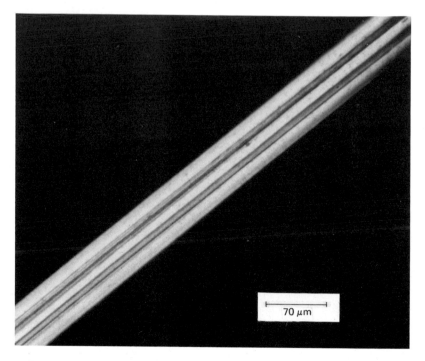

FIGURE 6-3. Longitudinal view of Antron® II nylon in NE–SW position of brightness, between crossed polars. The bands parallel to the axis of the fiber represent 4 tubes. In the N–S and E–W positions, the fiber exhibits complete extinction. Photomicrograph by I. A. Morrozoff.

FIGURE 6-4. Dual-phase, commercial acrylic fibers in cross section. One phase is highly pigmented and the other is not. Smallest division of the stage micrometer scale (inserted image at equivalent magnification) is 0.01 mm, i.e., 10 μm. Cross sectioning and photomicrography by J. J. Clark, American Cyanamid Co.

stiffness. Other fibers, such as cotton, sometimes are impregnated with a resin of the melamine-formaldehyde type to give stiffness to fabrics like buckram and crinoline. Various resins also are used to impregnate or coat fibers to give crease or soil resistance, dyeability, or inkability to cloth and carpeting.

Plastic hose and pipe often are coated to increase their resistance to solvents, weather, sunshine, or underground water. The pipe may be constructed like a tire (Figure 2-2) (Chapter 2), with cords or fabric to give strength and puncture resistance.

Fragments and *fibers* of vegetable or animal origin have long been used as reinforcing agents to improve the impact strength of phenolic, urea, and melamine resins. One of the first (and still a favorite) was wood flour, illustrated in Figure 6-5. Characteristic wood cells are shown filled and surrounded by phenol-formaldehyde resin. Other useful fragments are ground cork, ground coconut or walnut shells, ground peach pits, and crushed cornstalks or sugar cane. All of these are brown, a color which is acceptable in brown phenolic resins. But urea and melamine resins are colorless, and to keep the color neutral, colorless alpha cellulose floc is used. It is derived from cotton linters, wood pulp, or other sources of cellulose by getting rid of the brown lignin through paper-making techniques.

Man-made fibers also are being used to reinforce resins. For example, Figure 6-6 illustrates the shape, size, and distribution of acrylic fibers in a commercial diallylphthalate polymer. The widespread use of glass fibers (because of their exceptional strength) has already been described.

Uncrystallized phases may be crystallizable or uncrystallizable. A crystallizable phase may be in the glassy or rubbery phase, depending on the temperature. In Figure 9-3 (Chapter 9), an experimental system composed of 50% isotactic crystallizable polystyrene and 50%

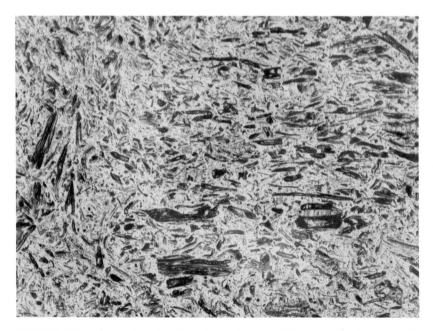

FIGURE 6-5. Wood flour in phenolic resin, in polished section by vertically reflected light.

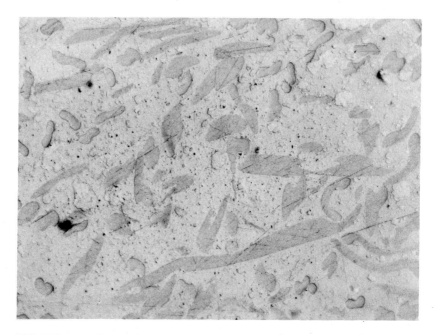

FIGURE 6-6. Photomicrograph by reflected light of a polished section of a molding of commercial diallylphthalate polymer filled with acrylic fibers.

atactic, uncrystallizable component is shown. The crystallized portion represents only about 10% of the whole system; not much is shown in the photomicrograph. That is, much of the crystallizable had not yet crystallized. The crystallized areas are seen to be composed of small, unicrystalline units. These areas are an integrated part of the whole, as illustrated by a crack running right through one of the areas.

Some multiphase systems for three-dimensional articles are sometimes called *polyblends* ("gum" plastics or "alloys"). A common type is one of elastomeric particles dispersed in polystyrene (Figure 5-3). Similar types are polyacrylonitrile-butadiene-styrene (ABS) and polystyrene-acrylonitrile (SAN).[2] The purpose of having the dispersed phase is to impart more impact strength to the polystyrene. The dispersed phase in each of these is generally so finely divided that an electron microscope may be needed to resolve it adequately.

Other physical mixtures may utilize larger resinous particles which differ in refractive index, color, reflectivity, hardness, or some other property. Moldable pellets of poly(methyl methacrylate) mixed with those of a terpolymer of methyl methacrylate produced the optical effect shown in Figure 6-7. The schlieren effect is manifested by a difference in refractive index of 0.025. The left-hand part of the figure shows that when a panel molded from a mixture having this difference in refractive index is placed close to an object, the object can be seen sharply, but when such a panel is placed some distance from the object, the observed image is diffuse. These and other optical properties suggest practical and decorative uses[3] of the old phenomenon of schlieren.

Parts of a single phase may differ subtly, yet be perceptible and useful. For example, one part of a glassy or rubbery phase may be under more stress than another. A stretched rubber band is stiffer than an unstretched one, and in this sense it is stronger. The explanation is that the rubber becomes oriented, and the orientation is manifested by a double refraction that is perceptible between crossed polars. Double refraction (birefringence) is also exhibited by resins and polymers in an effective state of stress (Figure 6-8).

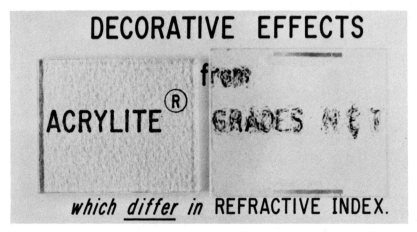

FIGURE 6-7. A transparent acrylic resin, pressure-molded from an equal mixture of two grades, differing in refractive index by 0.025. Left: the sample placed directly on top of the sign. Right: the sample raised somewhat above the sign.

Pigment particles in large proportion render a resin or man-made polymer opaque, but such particles in much smaller proportion render the same polymer translucent, or, in the case of man-made fibers, de-lustered (Figure 6-9). Pigments also have other effects: sometimes they act as nuclei for crystallization, and sometimes they inhibit crystallization, as demonstrated in Figure 6-10.

Organic resins, polymers and plastics lack the resistance to wear, abrasion, and fire exhibited by ceramic and similar inorganic materials. Hence inorganic powders, particularly inexpensive ones such as calcium carbonate and silica, find use in resins. In a related application, powdered glass waste is being tried instead of limestone in asphalt for pavements.[4] Pumice, ground quartz, and ground glass are

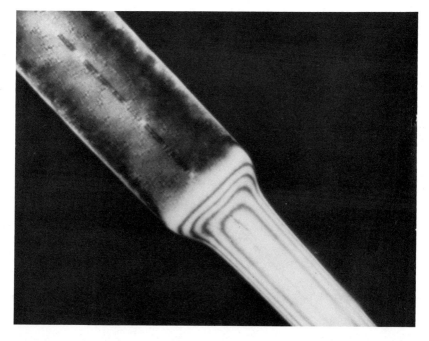

FIGURE 6-8.   Unstressed nylon (upper left, with tiny spherulites) necked down under stress (lower right with strain bands) between crossed polars.

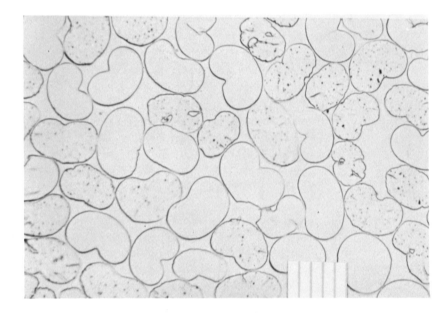

FIGURE 6-9. Acrylic commercial carpeting fibers in cross section. Some of the fibers have been delustered with pigment particles and some have not. Smallest division of the stage micrometer scale (inserted image at equivalent magnification) is 0.01 mm, i.e., 10 μm. Cross sectioning and photomicrography by J. J. Clark, American Cyanamid Co.

used in polyester writing boards for classrooms. Powdered quartz is being used in resinous dental fillings. Fly ash is the solidified molten spray produced by the burning of powdered coal in power plants. It is composed chiefly of tiny spheres of fused silicious ash, and all of the spheres are characteristically uniform. It is available in huge quantities, and presents a ready-made filler for some types of plastic.

Some other inorganic fillers are of an opposite type: they are abrasive because they are hard and have sharp edges. In one example, phenolic resins and rubber are used as binders in grinding and cutting wheels, which contain corundum or silicon carbide abrasives.

Metallic particles are used in some resins to impart electrical conduction or heat conduction, as described in Chapter 7.

FIGURE 6-10. Nylon 6 coarse filament in cross section showing high degree of spherulitic crystallization on the left (0.1% TiO$_2$) and low degree on the right (3.2% TiO$_2$). Photomicrographs submitted by G. E. Coven, American Cyanamid Co.

The structure and morphology of artificial *sponges* and rigid *foams* are described on Level IV because they have at least two phases: the cell wall and the cell content. The structure of the cell walls is polyhedric for close packing, especially the pentagonal dodecahedron.[5] Ideally, the morphology of the cell should be equant, but unequal pressures during frothing may cause some elongations of the cell.[6] Unlike long crystals, however, elongated foam cells have their face angles changed.

The size of the cell is very important. Smaller cells develop less elongation in the direction of the rising foam. Although cellular plastics occasionally contain less than 60 cell units per cubic centimeter, 20,000–200,000 are more typical numbers.[7] Cell sizes are generally distributed statistically in a given product, with standard deviations from about ⅛ to ¼ (occasionally to ½) of the average unit's volume. Exact averages and ranges seem to depend on both the system foamed and on the manufacturing process.[8] A small proportion of special silicone fluid serves as surfactant* to control the cell size.

Latex (water-base) paints consist of tiny particles of soft resin and pigment dispersed in water as a stable, thick suspension. The principal

*"Surfactant" is short for surface-active agent.

part of the resinous phase is poly(methyl methacrylate) for outdoor house paints, and usually poly(vinyl acetate) for indoor paints. Automobiles, refrigerators, and other steel products are now being coated with resinous finishes by an electrostatic mechanism[9] whereby a fine powder of pigmented resin is first deposited as a layer of controlled thickness and then is fused in place. This avoids discharge of solvent vapors into the air, and even avoids the water and emulsifiers of latex paints.

Natural mineral fibers (commonly and loosely called *asbestos*) are of several species, including chrysotile (most common), crocidolite (blue; chiefly used for its chemical inertness and for its stability in combustion reactions), tremolite, and anthophyllite. Chrysotile is a serpentine, which is a hydrated magnesium silicate. This means that water is lost from the composition when it is heated above 600°C, thereby destroying the structure. Nevertheless, it is by far the most important asbestos. The most important *man-made inorganic fibers* are of silicate glass, but some are of fused silica. "Whiskers" of boron and of various metals, grown by thermal decomposition of gases, are now available and find their way into resinous products intended for electrical, thermal, and magnetic applications. Similar whiskers of graphite[10] are used to impart increased strength to jet engine parts.

SUMMARY

Regardless of all the information which might have been obtained about a material on the levels of its composition, its molecular structure, its phases and its surfaces, the designer and the applications engineer must know its bulk properties. How strong is it? How dense? At what temperature does it soften or weaken? How does it stand up to sunshine and weather? What are its electrical properties? Any potential user of the material must have precise and dependable quantitative answers to these questions before he can design and manufacture a product. To obtain these vital data he turns to the standardized tests developed by producers and users together under the guidance of the American Society for Testing and Materials (ASTM). Dependable procedures have been worked out for testing all kinds of behavior, and these procedures are constantly being refined as instruments and techniques are improved. The student of resinography must become familiar with the ASTM, and must know how to find what he wants within its vast battery of standard tests.

The study of pure materials themselves in terms of composition, structure, phases, and surface behavior is deficient in still another way: pure materials are often used in conjunction with other pure materials to obtain important special advantages, and so it is the *composite* which must be studied on Level IV. Composites are polyphase materials in which some suitable matrix is reinforced by carefully chosen fibers or powders or other substances, or the matrix is coated with something else to improve its performance. Many examples are considered in this chapter: wood and bone among the natural composites, steel and precipitation-hardened aluminum among the polyphase metals, resins made much stronger and resilient by incorporating coated glass fibers, and rubber reinforced with carbon black. *Laminates* are another class of composites: paper is laminated with resins, thin wood plies are laminated with adhesives, and hose and tires are constructed in layers, each layer having a special purpose.

Still another class of polyphase materials utilizes suspensions of one material in another of closely related composition. A crystallizable polymer may be only partially crystallized, resulting in crystallites of the polymer embedded in an amorphous matrix. One polymer may be dispersed in another to produce a polyblend, or an elastomer may be dispersed in a stiff polymer to improve its impact strength. The object of such combinations may not be just improvement of mechanical properties; sometimes a marked change in optical properties is achieved, as when pellets of different clear plastics are molded together to produce schlieren effects (resulting in high light transmission but diffuse images at a distance).

Foams and sponges illustrate still another aspect of composite materials. Here a gas expands a polymeric or elastomeric matrix, forming tiny gas cells of some optimum size controlled by a silicone surfactant. When there is a uniform distribution of forces in the mass, the gas cells form close-packed polyhedra, but in industrial products the cells are usually elongated in the direction of expansion of the foam.

Almost all of the principles of composite materials are illustrated in a paint film. Here the fluid vehicle becomes a matrix for the particles of pigment and filler, and also an adhesive to secure the film in place. All the known agents and devices are used to achieve and maintain dispersion of the solids in the liquid paint, but this is temporary in that some segregation or layering often is desirable in the finished film. Testing the final film for durability and for resistance to weathering agents involves painstaking procedures which are spelled out in the ASTM tests.

SUGGESTIONS FOR FURTHER READING

S. Minami, Morphology and mechanical properties of polyacrylonitrile fibers, in *Acrylonitrile in Macromolecules* (Eli Pearce, ed.), pp. 145–157, ACS Applied Polymer Symposium No. 25, John Wiley and Sons, Inc., New York, N.Y. 10016 (1974).

M. C. Botty, C. D. Felton, and R. E. Anderson, Application of microscopical techniques to the evaluation of experimental fibers, *Textile Research J.* **30**, 959–965 (1960).

L. Boor, Hardness, abrasion, and wear resistance testing of plastics, *ASTM Bulletin* **244**, 43–47 (Feb. 1960). L. Boor, M. Hanok, F. S. Conant, and W. E. Scoville, Jr., Rheological testing of elastomers at low temperatures, part II, *ASTM Bulletin* **246**, 25–32 (May 1960). F. W. Reinhart, C. Brown, L. Boor, and J. J. Lamb, Evaluation of the Boor-quartermaster snag tester for coated fabrics and plastic films, *ASTM Bulletin* **210**, 50–59 (Dec. 1955). L. Boor and S. L. Trucker, An improved fadeometer, *ASTM Bulletin* **189**, 38–43 (Apr. 1953). L. Boor, Some notes on the effect of testing machines on the tensile properties of plastic films, *ASTM Bulletin*, 47–53 (Dec. 1949). L. Boor, J. D. Ryan, M. E. Marks, and W. F. Bartoe, Hardness and abrasion resistance of plastics, *ASTM Bulletin*, 68–73 (Mar. 1947). L. Boor, Indentation hardness of plastics, *Proceedings of the American Society for Testing and Materials* **47**, 969–991 (1944). B. Maxwell, Hardness testing of plastics, *Modern plastics*, Breskin Publications, Inc., New York, N.Y. 10022 (May 1955).

REFERENCES

1. ASTM designated standards are in any current *Annual Book of ASTM Standards* under the part number given in the annual *Index to ASTM Standards*, American Society for Testing and Materials, Philadelphia, Pa. 19103.
2. ASTM designation D1600, Standard abbreviations of terms relating to plastics, *ASTM Standards*, part number given in *Index*, printed annually by the American Society for Testing and Materials, Philadelphia, Pa. 19103.
3. T. G. Rochow, assigner to American Cyanamid Co. of U.S. Patent 3,345,239, Oct. 3, 1967, Method for producing decorative articles of manufacture, U.S. Patent Office, Washington, D.C. 20231.
4. W. G. Mullen, Skid resistance and wear properties of aggregates for paving mixtures, *Project ERSD-110-69-1, Highway Research Program*, N.C. State University, P.O. Box 5993, Raleigh, N.C. 27607 (1972).
5. E. A. Blair, Cell structure and physical properties of elastomeric cellular plastics, in *Resinography of Cellular Plastics* (R. E. Wright, ed.), ASTM Special Technical Publication 414, pp. 84–95, American Society for Testing and Materials, Philadelphia, Pa. 19103, (1967).
6. R. H. Harding, morphologies of cellular materials, in *Resinography of Cellular Plastics* (R. E. Wright, ed.), ASTM Special Technical Publication 414, pp. 3–18, American Society for Testing and Materials, Philadelphia, Pa. 19103 (1967).
7. G. F. Baumann, S. J. Aulabaugh, and S. Steingiser, Investigation of urethane foam formation by motion-picture photography, in *Resinography of Cellular Plastics* (R. E. Wright, ed.), ASTM Special Technical Publication 414, 68–83, American Society for Testing and Materials, Philadelphia, Pa. 19103 (1967).
8. D. Cannell, Paint, in *Kirk-Othmer Encyclopedia of Chemical Technology,* 2nd ed., Vol. 14, pp. 462–485, Interscience Div., John Wiley and Sons, Inc., New York, N.Y. 10016 (1967).
9. G. E. F. Brewer, Electrodeposition of Coatings, in *Encyclopedia of Polymer Science* (H. F. Mark, ed.), Vol. 15, pp. 178–191, Interscience Div., John Wiley and Sons, Inc., New York 10016 (1971).
10. Clauser, H. R., Advanced composite materials, *Scientific American* **229**, 36–44 (July 1973).

7

7

MORE ABOUT COMPOSITES

In the previous chapter it became evident that polyphase systems were the rule, rather than the exception, in engineering materials derived from resins, polymers, and elastomers. We saw why this was so, and we considered some examples of how plastics were strengthened or stiffened or bettered by the inclusion of powders, fibers, fabrics, and films. We turn now to some more elaborate examples of composite materials which come under the survey of the resinographer, examples which call for a more detailed description.

In the world of materials there are four "continents" (see Frontispiece), portraying different kinds of plastic materials which in the past have been utilized quite separately in arts and crafts, in construction, and in industry. Any two or more of these kinds of materials may be combined into a composite material. Until recently a composite was regarded simply as a linear combination of the constituent materials, but the resulting combination may have its own characteristics that oftentimes cannot be calculated directly or completely from the properties of the separate materials. The result is synergism, where the total becomes something more than the sum of its parts.

After the superior properties of several early composites were fully established and corroborated, specialists in technology attempted to explain the reasons for the synergism. Chemical engineers emphasized that the strength of chemical bonding between the different phases had been underestimated.[1] Metallurgists thought the role of structure, especially microstructure, might have been underestimated. Physicists studied x-ray diffraction patterns of the phases and then, by electron microscopy and field-emission microscopy, brought in the role of imperfections to explain deformation and the nature of grain boundaries. Finally, after the extreme demands on materials brought about by the space age, the physicist, chemist, and engineer have had to pool their ideas and techniques in the new area of *materials science*. Today, a unified understanding of materials seems even more important as we strive to make the best use of diminishing resources and to re-use materials that had formerly become only embarrassing waste. So we find resinography taking on a greater importance as it becomes integrated with the other disciplines of materials science.

Resinography has never been independent of metallography or petrography, nor separated from the study of ceramics and waxes. In his concern with all aspects of the materials he studies, the resinographer has to be ready to call on all the related disciplines in order to complete his study. The following examples will illustrate this point.

COMPOSITES WITH METALS

Tires such as those of the old clincher type have long been composites of rubber, fibers, and steel. The antique tires shown in cross section in Figure 7-1 are of this construction. The "beads" of each tire contain a cable of steel wires. In most cases the stiff, parallel wires were wrapped together spirally with a pliable wire. Specimens of the two kinds of wire are shown metallographically in Figures 7-2 and 7-3. The binding wire (Figure 7-2) is seen to be of very low-carbon steel, mechanically worked (drawn) and then annealed to make it malleable again. In longitudinal view (left), the grains or unit crystals are slightly elongated in the direction of the wire. In cross section (right), the grains nearer the periphery of the wire (lower left) are larger than those

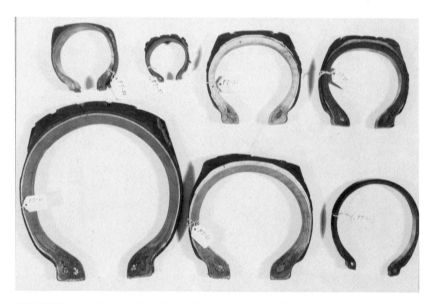

FIGURE 7-1. Some antique tires. Almost-vertical, reflected light was used. Bead wires were cut with a hacksaw, and the rest of the tire was cut with a knife. Wires were examined metallographically in cross and longitudinal sections. Cords were examined by transmitted light after mounting in oil. Tread stock was examined resinographically.

nearer the center (upper right), indicating that before annealing, the peripheral part of the wire received more cold-working during drawing than the center of the wire. Figure 7-3 shows views of a typical bead wire. Evidently the wire had been left severely drawn to make it stiff and strong. Both views show that the structure is that of a bundle of fine metallic *fibers*, thus telling why the wires have their present properties.

Figure 7-4 shows the cross section of an adhesive bond between two sheets of aluminum alloy, as polished for examination. Before coating each sheet with phenolic varnish (two dark layers), each aluminum sheet had been anodized (dark dots) to make the phenolic

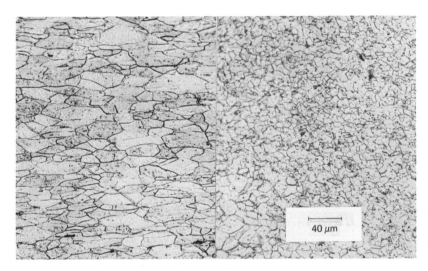

FIGURE 7-2. Annealed wire from an antique tire; binding wire. Vertical reflected light was used. Left: longitudinal section. Right: cross section. Large grains are near the periphery, small grains in the core.

FIGURE 7-3. Cold-drawn wire from an antique tire; bead wire. Left: longitudinal section. Right: cross section.

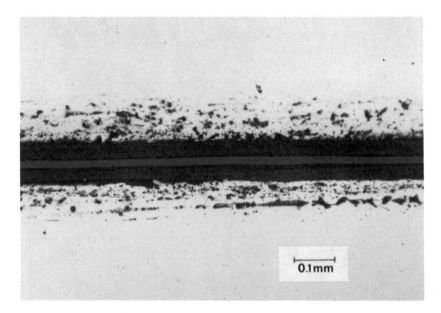

FIGURE 7-4. Resinous bond for aluminum alloy. The white areas, top and bottom, are aluminum alloy; the central, medium gray layer is elastomer; the two dark gray layers are varnish-primer.

resin stick better.* The central layer (medium gray) is an elastomer to give some more resilience to the joint. The results of tensile tests on this composite were good, but the costs of so complicated a sandwich structure were much higher than the cost of using a composite, one-step "spread" (Figure 7-5). The continuous phase is a phenolic varnish and the particulate phase is an elastomer. Another advantage of the particulate system is that it is three-dimensional in application and properties.

*The layer of aluminum oxide built up by anodic oxidation adheres tenaciously to the metal because of *very* strong bonds between aluminum and oxygen atoms (the heat of formation of Al_2O_3 is 390 kcal/mole) and because there is no demarcation between oxide and metal. Extra aluminum is present in the adjacent oxide, and there is every graduation from metal to oxide. At the other side of the layer, the hydrated aluminum oxide exhibits its basic properties and combines with acidic OH groups in the phenolic resin, establishing a strong bond.

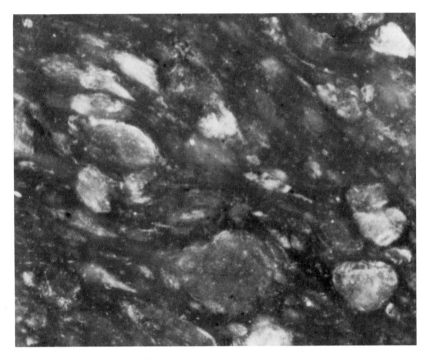

FIGURE 7-5. Mixture or "spread" for one-step bonding of aluminum sheets. Continuous phase is phenolic varnish; particulate phase is an elastomer.

Particulate composites have their own advantages and limitations. An interesting example is illustrated in Figures 7-6 and 7-7, which are polished cross sections of a good (Figure 7-6) and a poor (Figure 7-7) commutator brush for electric generators. The electric current is carried chiefly by the contiguous particles of pure copper (white). Lubrication is furnished by the graphite particles (light gray). Adhesion and consolidation are provided by the continuous phase of melamine-formaldehyde resin (dark gray). The black areas represent voids. Their population, size and distribution represent one of the differences between the good and poor brushes. The sizes and shapes of the copper particles also are different. In this and in every case of quality control, empirical results can be interpreted, evaluated, and then tested by experiment, using resinographic techniques.

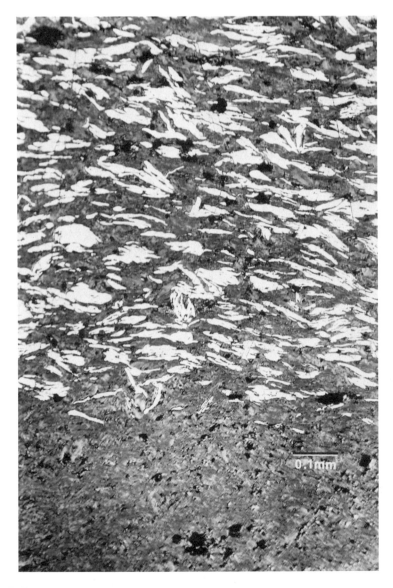

FIGURE 7-6   Polished cross section of a *good* commutator brush for electric generators. White = copper (for conductivity); light gray = graphite (for lubrication); dark gray = melamine-formaldehyde resin (consolidation); black = air (pores).

FIGURE 7-7. Polished cross section of a *poor* commutator brush for electric generators. White = copper (for conductivity); light gray = graphite (for lubrication); dark gray = melamine-formaldehyde resin (consolidation); black = air (pores).

COMPOSITES WITH GLASS, MINERALS, AND METAL OXIDES

Inorganic glass is a *thermoplastic*, and therefore might be included directly in resinography, but the composition, treatment, properties, and behavior of glass traditionally are taken up in the study of ceramics,[2,3] not resinography. Therefore the subject of glass is introduced here only in connection with its use as a reinforcing agent in resinous composites.

Glass is commonly used with resins in the form of fibers thin enough to be flexible. Such fibers (continuous filaments) are spun into yarns to make glass cloth. The cloth is then impregnated with a resinous polymer and pressed between heated platens to give a distinctive material (Figure 7-8 shows a polished cross section of such a material). What Figure 7-8 does *not* show is that a great deal more must be done in order to make a really good composite. As received, the glass cloth is contaminated by a starchy or oily sizing that is put on the glass fibers so that they will not scratch each other during the spinning and weaving processes. This sizing is inferior in thermal stability and moisture resistance to the phenolic or epoxy resins which will later be used as matrix, so a good bond between the glass fibers and the polymeric matrix would not be maintained. Some improvement can be made by burn-

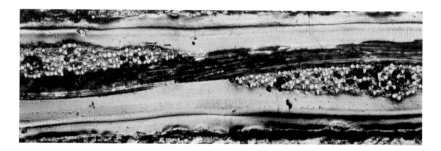

FIGURE 7-8   Glass cloth impregnated with a resin, polished cross section. The glass cloth manifests a simple basket weave. The white dots are ends of glass fibers in a warp thread. Filler thread is shown going under left warp thread and over right warp thread. The smooth matrix is resin.

ing off the sizing (heating the glass cloth in air at 500°C until white) before impregnating with the resin, but the bond still would not be good because the surfaces of glass and resin would not be sympathetic in the chemical sense. When a composite of cleaned glass cloth and phenolic resin is boiled in water, the water seeps in along the glass surfaces and soon weakens the composite. The remedy is to treat the cleaned glass surface with a reactive organosilicon "coupling agent," which bonds itself to the glass by Si—O—Si bonds and presents an organic surface to the resin for proper wetting and adhesion. After such treatment, a strong, durable composite is obtained.

Various minerals, natural or modified, are used in many plastic products.* A good example is the abrasive constituent in chalkboards for classroom use. The old-fashioned blackboard of natural slate is naturally abrasive to chalk, but the deposits of suitable slate are almost exhausted. Man-made writing boards are generally made of colored laminated or coated plywood or hardboard, treated on the surface to receive chalk by gentle abrasion. One class of abrasives tried in writing boards coated with polyester resin was known as "silica flour,"[4] without any specification of structure (amorphous SiO_2, quartz, silica glass, or whatever) or of morphology (size-range, average size, or shape of particles). Figures 7-9 and 7-10 show two distinct morphological types of quartz. Figure 7-9 shows powdered massive quartzite, which yielded angular fragments varying greatly in size. Figure 7-10 shows powdered Arkansas novaculite, which originally was a fine sandstone of relatively round, uniform quartz grains, which were originally cemented together but were effectively separated by a particular kind of mill. Obviously the two powders will have very different abrasive qualities.

Many other mineral and ceramic fillers have distinctive morphological features. Pumice (volcanic glass) is generally porous, so its fragments are angular and not orientable. Vermiculite, like all micas, comes apart in flakes, and so its particles are readily oriented to lie in a plane. "Essence of pearl," while not a mineral (it comes from fish scales), is composed of flat crystals which are readily oriented in a

*See the discussion of asbestos on p. 119 in Chapter 6.

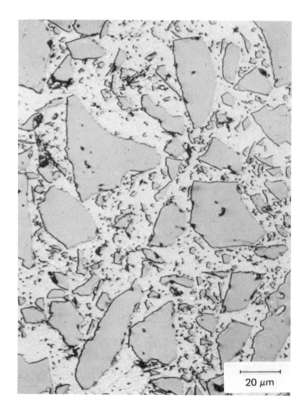

FIGURE 7-9.   Silica flour by reflected light (powdered massive quartz).

resin such as melamine-formaldehyde to make pearly buttons. Man-made pearlescence is simulated by other transparent, thin, *flat* (orientable) particles which differ sufficiently from the resin in refractive index.[5] Other types of oriented lustering agents are being simulated by designing and controlling the orientation of thin flat particles in a resinous matrix. Again both morphology (crystal habit) and structure (refractive index) are important.

It is essential that particles of a good white *pigment* be uniformly small and of high average refractive index. Titanium dioxide has surpassed all other substances as a white pigment for paints and plastics,

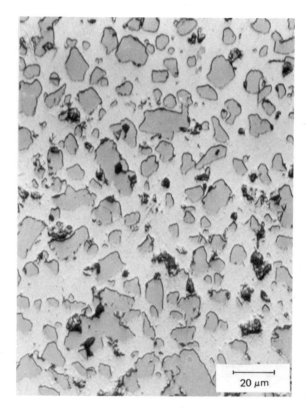

FIGURE 7-10.   Powdered Arkansas novaculite.

and as a "delustering agent" in synthetic fibers, because of its covering power (high average refractive index). It can be made in two structures, rutile and anatase. Rutile, with its higher refractive indices, has
the greater covering power, but in the dried paint film anatase
"chalks," i.e., comes to the surface of the weathering film to give a
renewed whiteness. So outdoor paints usually contain a mixture of the
two phases of TiO_2, controlled in manufacture by seeding some
anatase into the precipitating bath. The two different phases are recognized by their differences in crystal structure.

COMPOSITES WITH WAXES

The familiar wax paper, a thin tissue paper impregnated and sur-faced with hard paraffin wax, is a waterproof and greaseproof compos-ite that has done a good job for a hundred years. Stronger paper, laminated in several layers with oxidized paraffin or a similar sticky wax, serves much the same purpose in shipping containers. Waxes also provide a useful, final layer on a varnished or painted surface, simply because the wax layer is waterproof, wear-resistant, and re-newable. The common practice of waxing automobiles and furniture provides the best examples and the best rationale. For the sake of durability, hard waxes such as carnauba (a vegetable wax consisting of saturated esters with 47–65 carbon atoms) are used. Since the shine is produced by rubbing or by buffing, and the hard wax crystals resist spreading, much effort is required unless some silicone oil is added to lubricate the wax crystals during the buffing operation. In a typical polish, the dispersed crystals of wax and the emulsified globules of silicone oil (of refractive index 1.405) can readily be identified mi-croscopically in their aqueous medium.

SUMMARY

This chapter delves further into the synergistic combinations of plastic materials with metals, with minerals, with glass fibers, with manufactured pigments, and with waxes. It shows that materials from any of the four ''continents'' shown in the frontispiece may be com-bined with each other in a suitably imaginative and logical way to pro-duce composites with improved performance. The theory and practice of creating new composites comes under the new heading of materials science.

In the study of composites, the resinographer may find it neces-sary to use the related techniques of metallography and ceramic sci-ence in order to gain a complete description of the combined phases.

The necessity for knowing some metallography was illustrated by the study of tire constituents, bonded aluminum structures, and commutator brushes. In the study of glass-fiber composites, some knowledge of the drawing and sizing of glass fibers is needed in order to alter the situation and obtain a strong, water-resistant bond between the glass and the surrounding organic resin. Some facility in the microscopical examination of mineral fillers and inorganic pigments is also helpful in dealing with paints, coatings, and pigmented plastics. The examples in this chapter were chosen to illustrate techniques, and by no means cover the immense range of possibilities in analyzing composite materials.

SUGGESTIONS FOR FURTHER READING

Composite Materials, a comprehensive treatise in 8 volumes edited by L. J. Broutman and R. H. Krock, Academic Press, New York, N.Y. 10003 (1974). Volume 1 (A. G. Metcalfe, ed.) treats interfaces in metal-matrix composites. Volume 2 (G. P. Sendeckyj, ed.) treats the mechanics of composite materials. Volume 3 (B. R. Noton, ed.) treats engineering applications of composites. Volume 4 (K. G. Kreider, ed.) treats composites with metallic matrices. Volume 5 (L. J. Broutman, ed.) treats fracture and fatigue in composites. Volume 6 (E. P. Plueddeman, ed.) treats interfaces in composites with polymer matrices. Volumes 7 and 8 treat structural design and analysis, in two parts. This is the most comprehensive work on composites yet to appear.

REFERENCES

1. A. Kelly, The nature of composite materials, in *Materials: A Scientific American Book*, pp. 96–110, W. H. Freeman Co., San Francisco, Calif. 94104 (1967).
2. F. H. Norton, *Elements of Ceramics, 2nd ed.* Addison-Wesley Publishing Co., Reading, Mass. 01867 (1974). See also *McGraw-Hill Encyclopedia of Science and*

Technology, 3rd ed. Vol. 2 pp. 685–688, McGraw-Hill Book Company, New York, N.Y.10020 (1971).

3. C. E. Gracias and J. A. O'Dell, Applications of microscopy to the analysis of composite materials, *The Microscope* **17**, 161-167 (July 1969).

4. T. G. Rochow and R. L. Gilbert, Resinography, in *Protective and Decorative Coatings* (J. J. Mattiello, ed.), Vol. 5, p. 577, John Wiley and Sons, Inc, New York, N.Y. 10016 (1946).

5. H. A. Miller and L. M. Greenstein, Nacreous pigments and their optical effects, *Papers Presented at the New York Meeting of the ACS*, Division of Organic Coatings and Plastics Chemistry, pp. 424–436, American Chemical Society, Washington, D.C. 20036 (1963).

STATIC TESTING AND RECORDING

The engineer generally thinks of the practical, quantitative properties of materials as single, approximate values such as those found in a handbook[1] or encyclopedia.[2] Such values represent selected or average results culled from the technical literature, where the data come from standardized tests, especially those of the ASTM.[3] The tests are generally performed on comparatively large specimens, although some so-called microspecimens can be used when there is a severely limited amount of material.[4] A standard test piece is prepared as specified, and the piece is subjected to the pertinent forces or influences under standardized conditions, producing a number which is then widely accepted.

Let us take the practical property of hardness as one example. The engineer is accustomed to giving or taking one average value for hardness on a quite arbitrary scale.[5] Yet hardness is a rather vague term. It is described in mineralogy on the basis of 10 numbers, from 1 for talc to 10 for diamond. This Mohs' hardness number, as it is called, depends on what standard mineral just barely scratches another.[5] Rough as it is, with unequal increments, Mohs' hardness scale is still with us and serves as a first basis of classification.

The metallurgist is accustomed to describing hardness by measuring the dimensions of the depression made by a Rockwell steel ball or a Knoop diamond point under a standard load.[6] *Macro*-impression testing for hardness, using a ball indenter, has been tried on plastics,[5] but many plastics recover before the impression can be measured after the release of load. For this reason, the Knoop diamond point has been used. This makes a long rhomb-shaped depression in the sample. The long diagonal is "practically" unaffected by elastic recovery, whereas the short diagonal recovers quickly. Thus two simple numbers represent approximately what happened during loading and unloading of the Knoop indenter.

The Bierbaum *micro*hardness tester[5] drags a diamond point under a standard load across the various phases of a surface, producing a scratch. Microhardness numbers are calculated from the width of the scratch (in μm) formed at a certain load (3–9 grams). Scratch hardness would seem to be more closely related to suitability of a plastic for tableware, let us say, than impression hardness. Neither test can be observed while the deformation is taking place, but the measurement can be made immediately afterward. Moreover, the hardness of each kind of phase can be observed microscopically. This kind of static approach to dynamic phenomena is very useful in description, provided that the limitations are understood and remembered. Often they are not.

For all their faults, the standard static methods of testing carry a time-honored respectability not to be ignored. In this chapter we shall consider some of the concepts and pitfalls involved in such testing, along with some methods for postmortem examination (and photographic recording) of the test specimens.

SAMPLING

The resinographical detective is no better than the evidence he receives or obtains. Here his attitude, experience, imagination, and all the attributes of his associates contribute very much to reliability. First, no matter how many or how few samples, are the important ones present? If not, can they be obtained? Can he or others visit the scene to observe the process? If not, should samples be synthesized, or should the process be simulated? Visits, syntheses, pilot processing, and dynamic approaches in general can be expensive. Money saved by the resinographer's ability to do with what he has, or can get economically, should go to his credit and reputation. However, he should never draw general conclusions from a sample he knows to be inadequate.

PHOTOGRAPHS

There are two static kinds of photographs: the snapshot and the posed picture. Both kinds are like drawings or any other illustration. They can be representative or nonrepresentative; they can picture the object or picture something else. They can be touched up or retouched, for better or worse. A photograph represents the object under only *one* set of conditions for over 20 attributes of visibility (Chapter 1). Everything depends upon the integrity and experience of the photographer and the resinographer.

Since there can always be some lack of communication or some misunderstanding when others are brought in, the resinographer may choose to be his own photographer. If so, he has a great deal to learn about equipment, about the available photosensitive materials, and about development and processing. He must consider all the possible kinds of lighting, and their direction and intensity, in order to bring out the particular aspect he seeks to record. Is the result clear to others, even to the untrained observer, with a minimum of explanation? Is all extraneous and distracting material absent? Is the photograph properly identified?

A well-conceived and beautifully finished photograph can carry a great deal of conviction, and is worth more than the traditional thousand words. However, the resinographer must be prepared to take a dozen (or even a hundred) pictures in order to get the "right" one.

A TYPICAL "POSTMORTEM" EXAMINATION

A chemical engineer once made an experimental batch of polystyrene using a Ziegler type of catalyst, so as to obtain a very high proportion of isotactic stereoisomer in the polymeric product. To determine the proportion of isotactic polymer, he extracted his product by refluxing with a boiling ketone as prescribed in the literature. He sent samples of the soluble and insoluble fractions of polymer to an x-ray diffractionist, who found, as expected, that the ketone-insoluble fraction was crystalline and that the soluble fraction was amorphous. Then the engineer determined the melting point by watching for the temperature at which a few crystals slid down to the bottom of a melting-point tube. The "melting point" was close to that published, and was within the limitation of this approximate method. The engineer therefore made some test bars from each fraction and proceeded with his static testing. To his great disappointment, the strength values of the two bars were not at all as different as he expected! Then, and only then, did he go to a resinographer for a "postmortem."

The resinographer put the specimens between crossed polars, under a microscope. As illustrated in Figure 8-1, the "insoluble" portion contained some of the amorphous phase, and the "soluble" portion contained some of the crystalline phase. It was a simple examination, and took little time, but the result was obvious and unequivocal.

What has been learned from this "postmortem"? First of all, the extraction method is a blind one. There is no way of seeing or otherwise sensing when the extraction is completed. One simply boils the sample with the solvent for an arbitrary length of time and assumes that all the soluble material has been extracted. In fact, many more assumptions are made in this approach. It is assumed that all of the crystallizable (tactic) phase has been crystallized, without considering such questions as: How insoluble is the crystalline phase? What are the rates of dissolving and crystallizing? Is there a large enough difference in solubility between atactic and isotactic, uncrystallized polystyrene? Is there only one crystalline phase to be separated? If not, the mixture will probably have a lower melting point than either component. The melting-point tube method utilizes the liquid from the first crystals that melt as the vehicle which carries the rest of the solid down the melting-point tube at the end point. But since impurities lower the melting point, the highest melting temperature would be that required by the purest crystals. The most important question is: What was the proportion of crystals in the polystyrene bars tested for strength? The temperature and pressure required for molding the bars could have

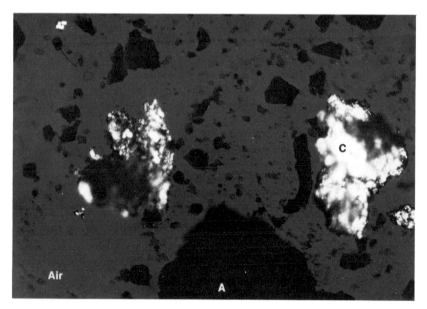

FIGURE 8-1. Polystyrene, sample V (ketone soluble), between almost-crossed polars (70×). By comparison with sample I, it was concluded that this sample contains some crystalline polystyrene (C) and is not "100% amorphous." Black (isotropic) particles are amorphous (A). Continuous space represents air.

changed the original proportions. Unfortunately, the broken test bars had been discarded, so that a "postmortem" could not be performed on them. (Never discard pieces until the case is settled!) Rather than do everything over, a different investigation was started and a great deal more was learned from a *dynamic* study, by watching the whole system melt and recrystallize. This kind of study will be explained in the next chapter.

SUMMARY

There are two major ways of testing bulk samples of material to obtain values of its engineering properties. One way is to subject a standardized sample to appropriate forces applied in a standardized way, and to record the end result (tensile strength at break, dimensions of indentation from a hardness tester, or whatever) after the event is over. This is called *static* testing. It produces one datum, which is accepted all too often at its face value. The other general method is *dynamic* testing, in which continuous observations are made while an action is taking place. The aim here is to observe and understand just what happens while it is happening, and to notice especially the progressive changes that occur.

In this chapter, static testing is examined as a primary source of engineering information. The advantages are simplicity, speed, and an unequivocal numerical result, even in unskilled hands. The chief disadvantages are that there is no monitoring, and the end result teaches little about what changes were taking place during the testing action. To illustrate this point, the various methods for measuring the common property of "hardness" were considered. The usual metallographic method, in which a small hardened steel ball is pressed into the sample under known force and then the diameter of the indentation is measured, was found unsuitable for plastics because most polymers show some recovery after a force is removed. Using a sharp-edged rhomb—shaped diamond point instead of a steel ball leaves a more permanent indentation, but involves some cutting of the surface. A better method for plastics involves scratch-testing, in which a loaded diamond point is dragged across the surface of the plastic, and the width of the scratch is measured microscopically afterward. By this method, the scratch hardness of each phase can be measured.

The results of static testing are no better than the sampling procedures used to obtain the specimens. How and where to obtain samples, as well as how many to take and how to select them, are matters of training and experience. The help of others often is necessary in getting the required samples, and the reliability of their efforts must be assessed as well.

Photography is a favored method of recording and presenting evidence obtained by testing, and it is very effective in the hands of a resinographer who knows how to arrange the lighting, how to choose and operate the equipment, and how to turn out a well-finished and properly-identified print. This job usually cannot be handed to others.

It requires skill, experience, judgment, and a great deal of patience to get the one "right" picture.

Standard static tests carry a time-honored respectability and so are widely accepted, but their faults should always be kept in mind. One of the worst faults is that the test usually is conducted blindly, with no monitoring to insure that the procedure is actually accomplishing efficiently or completely the operation for which it was designed. This shortcoming was illustrated by an experience in preparing and testing isotactic polystyrene. The procedure for separating the crystalline polymer from the amorphous polymer was followed to the letter, but when the two products were molded into bars and tested for strength, their performances were almost the same. Resinographic examination of the products showed that both contained crystalline and amorphous polymer. Without a way to monitor the procedure for separation, or a way to check the specimens before testing, the operation was conducted blindly and failed in its purpose.

REFERENCES

1. *Handbook of Chemistry and Physics, 55th ed.*, 1974–1975 (frequent editions). CRC Press, Chemical Rubber Co., Cleveland, O. 44128.
2. *Modern Plastics Encyclopedia*, annual issue of *Modern Plastics*, McGraw-Hill Book Co., New York, N.Y. 10020.
3. *Annual Book of ASTM Standards*, American Society for Testing and Materials, Philadelphia, Pa. 19103.
4. T. G. Rochow and R. J. Bates, A microscopical automated microdynamometer microtension tester, *ASTM Materials Research and Standards*, **12**, 27–30 (April 1972).
5. T. G. Rochow and R. L. Gilbert, Resinography, in *Protective and Decorative Coatings* (J. J. Mattiello, ed.), Vol. 5, pp. 583–586, John Wiley and Sons, Inc., New York, N.Y. 10016 (1946).
6. ASTM designation E384, Standard method of test for microhardness of materials, *Annual Book of ASTM Standards* and the annual *Index to ASTM Standards*, American Society for Testing and Materials, Philadelphia, Pa. 19103.

9

DYNAMIC TESTING AND RECORDING

The purpose of this chapter is to give some examples of testing dynamic phenomena and the recording of these phenomena *while they are happening*.

DYNAMIC OBSERVATION OF A POLYMER SYSTEM

Crystals *grow* on nutrients as though alive; a colony of crystals has its own characteristics. If we wish to understand the behavior of crystals as individuals or groups, we should study them in their environment, rather than carry them off to a postmortem examination.

In one series of "live" experiments, a fifty–fifty mixture of polystyrene particles (soluble versus insoluble in ketone) was watched while being warmed on a hot stage under a stereoscopic microscope. The location and temperature of the last-melting (purest) crystals were observed.[1,2] (A narrow range of melting temperatures indicates purity; a wide range indicates a variation in composition of a crystalline solid solution.)

Recrystallization had been carried through 8 cycles, at which time the specimen had the appearance shown in Figure 9-1. Practically all of the crystallizable polystyrene had been crystallized. The crystals are

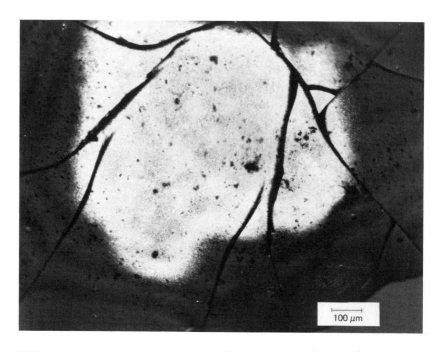

FIGURE 9-1.   Polystyrenes, 50% crystallizable: 50% uncrystallizable, incubated ca. 150°C for 5 hrs. The crystallizable portion is practically entirely crystallized. Sample was photographed between almost-crossed polars. (Small, dark particles are of a Ziegler catalyst.)

shown collected into one colony which was examined intangibly as such. Shrinkage cracks were seen to start in the crystalline region, showing it to be more brittle. The relative hardness could have been measured and observed continuously by making a straight (line-intercept) scratch with the Bierbaum tester described in the previous chapter. The whole crystalline region could also have been *tangibly* separated on the slide from the uncrystallized region. The resultant two fractions would then have been purer than the products of the blind, differential-solvent extraction described in Chapter 8. Further purification could have been obtained by repeating the fractional crystallization on the microscope slide. Then the purest crystals could have been examined not only for their principal refractive indices, but also for their many other optical and crystallographic characteristics,[3] to determine the presence or absence of polymorphism.

After melting the crystals shown in Figure 9-1, the melt was chilled to room temperature. Since there was no stirring, a composition gradient existed between crystallizable (isotactic) and noncrystallizable (atactic) regions. Within the composition gradient, cooling brought internal stresses which were relieved by cracking. When annealing was accomplished by reheating the crystalline mass, the cracks disappeared, as illustrated in Figure 9-2. There may also be seen some "islands" which were formed by the disintegration of a large island of glassy atactic material in the center, which was originally surrounded by a "sea" of rubbery atactic material. The islands now shown are the last of the glassy phase. The glass transition took place over a wide range of temperatures because of the atactic–tactic gradient. These temperatures were recorded, so that a phase diagram could be drawn. Figure 9-3 shows some colonies of crystals which reappeared during a

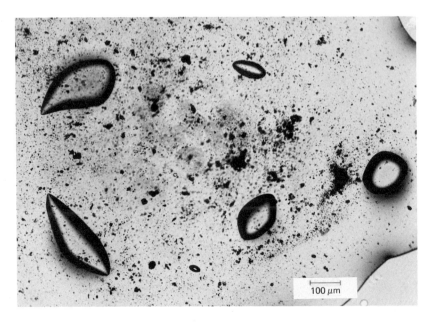

FIGURE 9-2.   Polystyrenes, 50% crystallizable: 50% uncrystallizable, reheated to 115.5°C. Photo was taken before cracking. Compare islands with those of Figure 9-1. Boundaries of remaining islands are black by total reflection of light at spaces differentiating volumes of old and new phases. (Small, dark particles are of a Ziegler catalyst.)

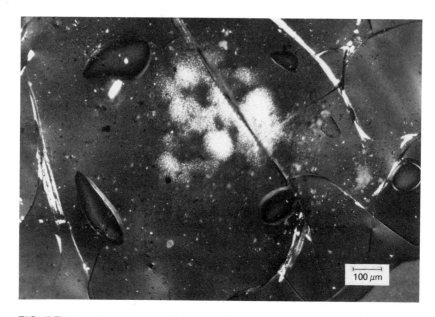

FIGURE 9-3. Polystyrenes, 50% crystallizable: 50% uncrystallizable, between almost-crossed polars. The melt was quenched in air, warmed until only islands of glassy phase remained, held at slightly lower temperature to obtain some colonies of crystals and was quenched again to freeze the three phases into coexistence with incidental cracks.

reheating cycle. Note that the cracks through the crystallizable region have become isotropic just prior to healing (annealing), whereas the cracks around the crystals and islands of glass are still anisotropic because of the remaining stress gradient.

OBSERVATION DURING DYNAMIC TESTING

Mechanical testing is dynamic, but only relatively recently has the motion been stopped to see what is going on. One example of a *macro*scopical continuous study is that of Wright and Hall.[4] They took high-speed motion pictures of standard notched specimens of resins during the Izod impact test. The typically brittle and ductile resins were observed between crossed polarizing sheets so as to show the changing photoelastic patterns during the impact test from the start to the breaking point.

The above example illustrates the bringing of a motion picture camera to a standard testing machine. As the relevant structure and morphology become smaller, the problem of bringing the observational apparatus to the tester becomes more difficult. Bringing the tester to the observation post can surmount this difficulty.

In the microscopical automated microtension tester,[5] a microdynamometer was built to fit on top of a microscopical hot stage, as shown in Figure 9-4. The microtension tester is operated by a small electric motor (M) which can be electronically controlled (Sp) to run at various constant speeds. The motor is geared (Gr) to a pair of micrometer screws (μ) which rotate in opposite directions to pull on both ends of the standard microspecimen. A small central hole is made with a hypodermic needle. The hole serves to concentrate the stresses. As the jaws are separated, the hole remains in the center of the microscopical object field because the specimen is pulled at both ends equally and simultaneously.

The micrometer spindles are nonrotating. They provide direct, firm action because one is fastened to the right-hand jaw and the other to the mounting of the transducer (D). The transducer is connected by wire to the electronic amplifier-recorder (K), which plots a curve of transducer output versus time.

The temperature of hot stage, H, is taken by means of a fine-wire thermocouple. The hot junction, which is placed near the specimen, cannot be seen, but its wires lead to the Dewar flask (A) and the potentiometer (P). The purpose of the glass tube (Gl) is to receive an air hose for bubbling air to stir and saturate the ice-water mixture around the cold junction in the Dewar flask. Figure 1-1 in Chapter 1 illustrates the typical correlation of load, time, and photoelastic effects.

Figure 9-5 shows four of the cinematographic frames close up. The typical photoelastic patterns are explained in the caption. When desired, motion pictures can be taken of the sample during testing to give a continuous dynamic record of the stress–strain relationship. Figure 9-6 shows one frame of such a moving-picture record of the stretching of a plastic film between crossed polars at 100°C, depicting

FIGURE 9-4. Microscopical automated microdynamometer together with speed controller (Sp), recorder (K), and other accessories.

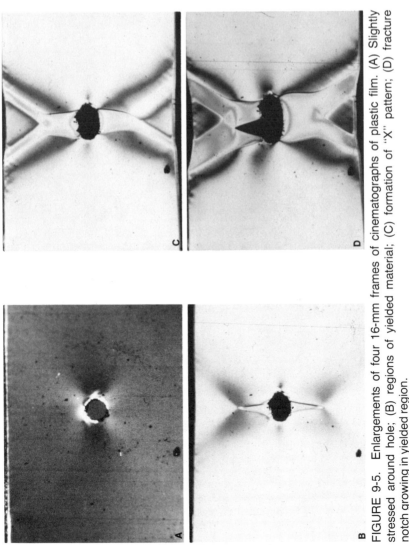

FIGURE 9-5. Enlargements of four 16-mm frames of cinematographs of plastic film. (A) Slightly stressed around hole; (B) regions of yielded material; (C) formation of "X" pattern; (D) fracture notch growing in yielded region.

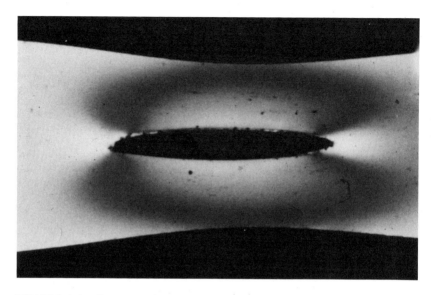

FIGURE 9-6. Enlargement of 16-mm cinematographic frame (ca. 12×). Stretching was done at 100°C.

the elongation of the center hole and the concentration of stress above and below it.

The advantages of a microdynamometer of the type shown in Figure 9-4 over the conventional instruments for dynamic testing are (1) it costs less to build; (2) there is continuous observation (both visual and photographic) of the specimen throughout the test; (3) there is built-in provision for time-lapse or motion pictures; and (4) polarization colors and patterns can be analyzed.

Another example of continuous microscopical study is that of Mogensen.[6] He took motion pictures while spinning a single filament of fiber-forming material from the orifice of a transparent glass cell which was fitted on a light microscope. The fiber-forming material was colored with Congo red and viewed with a polarizer which was divided so that one half had its direction of vibration at 90° to that of the other half. Thus the optical property of dichroism could be observed while the stream of material was being coagulated and taken up on rolls. Tension was applied by increasing the take-up speed until the filament broke, while the action was recorded by means of a high-speed motion-picture camera.

These examples illustrate the value of observing and recording dynamic changes more or less continuously. When this is done, the observer has not only a figure for the ultimate tensile strength of the material, but also a complete story of how the stresses develop and where the fracture or failure starts. The result is far more informative than static testing of the same amount of material.

SUMMARY

The advantages and disadvantages of static versus dynamic testing were considered in the previous chapter. The present chapter is devoted to some illustrative examples of dynamic testing. The first technique to be considered is continuous microscopical examination of the samples of the isotactic and atactic polystyrene described in Chapter 9, while the operations of recrystallization, melting, and freezing are carried out on the stage of a microscope. A second technique involves high-speed motion pictures of specimens of a polymer, under conditions of polarized light, while the specimens are subjected to an impact test. The third technique involves the continuous microscopical observation of a specimen while it is being subjected to tensile strength and elongation tests. Here an automated microdynamometer is described, a device that subjects a sample to symmetrical stretching while a continuous record is made by cinephotomicrography under crossed polars. The apparatus and its operation are described, and an example of the result is given in the form of four cinematographic frames. (Another record is shown in Figure 1-1, Chapter 1.) The advantages of such micro-dynamometer testing are (1) the apparatus is comparatively inexpensive; (2) there is continuous visual *and* photographic observation of the specimen throughout the test; (3) time-lapse or fast motion pictures can be taken; and (4) stress patterns within the specimen can be analyzed in terms of their polarization colors.

A somewhat similar continuous microscopical observation of the fiber-forming process is also described, in which a single fiber was spun from a glass cell fitted to a microscopical stage. These four examples point out the message: Continuous observation provides not only a figure for the ultimate tensile strength (or other property) of a material, but also a complete story of how the stresses develop and where the fracture or failure starts. Hence, more information is obtained from the same amount of material.

REFERENCES

1. T. G. Rochow, Resinography of high polymers, *Analytical Chem.* **33**, 1810–1816 (1961).
2. W. C. McCrone, *Fusion Methods in Chemical Microscopy,* Interscience Div., John Wiley and Sons, Inc., New York, N.Y. 10016 (1957).
3. E. M. Chamot and C. W. Mason, *Handbook of Chemical Microscopy*, 3rd ed., Vol. 1, John Wiley and Sons, Inc., New York, N.Y. 10016 (1958).
4. R. E. Wright and N. W. Hall, Transparent section microscopy and high–speed cinematography in photoelasticity studies, in *Resinographic Methods*, ASTM Special Technical Publication 348, pp. 17–30, American Society for Testing and Materials, Philadelphia, Pa. 19103 (1964).
5. T. G. Rochow and R. J. Bates, A microscopical automated microdynamometer microtension tester, *ASTM Materials Research and Standards,* **12**, No. 4, 27–30, 53 (1972).
6. A. O. Mogensen, Microscopical apparatus and techniques for observing the fiber-forming process, in *Resinographic Methods*, ASTM Special Technical Publication 348, pp. 31–35, American Society for Testing and Materials, Philadelphia, Pa. 19103 (1964).

SOME CONSTRUCTIVE APPLICATIONS OF RESINOGRAPHY

There are always two sides to science, the analytical and the synthetical. The analytical side strives to answer the questions How?, What?, and Why?, and it does so by taking apart the things or the processes under study so that we may learn about composition, structure, and interaction. The synthetical side starts with known substances and known reactions, and strives to create from them something new and useful. Previous chapters in this book have dealt with analytical investigations devoted to research, or to trouble-shooting in the plant, or to quality control of industrial products. In the present chapter, we shall turn to the other side and consider some constructive and creative aspects of resinography. The examples will be based on lessons learned earlier in structure and morphology, and will serve to illustrate the creative use that may be made of information gained in routine study.

POWDERS AND PREFORMS USED IN MOLDING

The favorite method for fabricating objects made of plastic is by *compression* molding. A permanent mold is prepared, a fixed quantity of solid molding material is poured in, and pressure is applied at a certain temperature for a certain length of time. The pieces or particles of solid material may be in the form of grains of powder, or as prepared pellets, or as flakes or fragments, according to the method of making them. Some of these methods are (1) casting fresh plastic composition or recycled scrap into large slabs, and crushing these to suitable fragments; (2) allowing drops of melted molding composition to solidify as beads or bubbles; (3) spray-drying a solution or suspension of the polymer to obtain a powder; (4) rolling out the soft plastic into sheets, and breaking these into suitable fragments; or (5) extruding a hot-melt in the form of a rod and then chopping the rod into pellets.

The fifth method (pelleting) is a common method of making particles out of thermoplastic compositions, because it is continuous and because the pellets are a convenient and dust-free medium for filling a mold. After the molding, the finished moldment may look homogeneous, but when it is annealed near but still below the glass transition temperature of the thermoplastic, a peculiar texture of the surface can be observed. As shown in Figure 5-12 (Chapter 5), there is a surfacial reticulation that is distinctly reminiscent of the morphology of the original particles. This is but one example of the well-known "memory"[1] of thermoplastic polymers. The novel aspect is that in this behavior we may find a way to texturize the surface of a pressed sheet or molded article. The surface in Figure 5-12 looks like that of a peened metal, but it is the surface of a molded and annealed sheet of plastic. The effect might be more realistic if a shiny metal, such as aluminum, were deposited upon the surface by vaporization in vacuum. Other artistic effects might be produced by spraying a colored lacquer at a low angle onto the surface. Still another variation would be to remold the "shadowed" surface to flatness or smoothness. The artist can produce still further variations at will by starting with single- or multicolored, transparent pellets.

A relevant patent[2] is based on varying the average refractive indices of two or more species of compatible pellets, either colorless or colored the same or different colors. The optical effects may also be diversified by varying the amount of flow (schlieren) of one type of pellet into the melt of another type. Obviously the extent of flow is controlled by the differences in softening temperatures among types of pellets.

One practical use derived from the foregoing resinographic observations (illustrated by Figure 10-1) would allow light to pass through windows efficiently, but would diffuse images seen from the outside. This could lead to new architectural designs of shower screens, room dividers, and exterior doors and windows.

The simulation of marble is an old trick of the molder who uses the flow patterns of differently colored pellets—transparent, translucent and/or opaque. The experienced observer can prescribe variations

FIGURE 10-1.   Part of an experimental plate with smooth plane surfaces placed over a piece of black paper to show the optical effect of compression-molding pellets of two different average refractive indices. (Shown actual size.)

in sizes and shapes of the pellets. These may be preformed for mass production and uniformity of the final product.

A great many other effects can be achieved by the same kind of compression molding, but it has the limitation of being slow, since a mold must be filled and emptied repeatedly. Injection molding is much faster.

In *injection molding*, the pellets or other particles are poured into a hopper which leads to a melting pot. From there, the melt is injected into a cluster of molds. For experimental purposes, standard[3] sizes and shapes are injection-molded and tested for comparison of properties with those of compression-molded objects of the same size and shape. During the development of higher impact strength in polystyrene by incorporation of an elastomeric constituent, streaks occasionally were found in injection moldments but not in their compression counterparts. Resinographic investigation showed that in severe cases a slightly pearly luster was present. This meant that there was a wavy structure of flakes or fibers within the injection-molded part. Experimentally the pearlescence was accentuated by varying differences in refractive index, compatibility, and color between two or more kinds of thermoplastic grains fed into the injection-molding machine. The resulting moldments (standard test bars) varied in appearance from mother-of-pearl to rock asbestos. The equant granules that were fed into the hopper had been changed to contiguous fibers, which must mean that the original pellets remembered their identity even though they were transformed into filaments (see Figure 10-2.) The moldments were very pretty to look at, and they were strong enough to be just as useful as their more homogeneous compression-molded counterparts.

The alert reader will be able to imagine situations where a fibrous structure deliberately introduced during injection molding could increase the lateral flex strength and durability of molded pieces for special purposes. He will understand also why fibers spun from a melt (really a process of injection molding through the holes of the spinnerette) have a helpful skin-and-core structure and a lateral anisotropy (see Figure 10-3).

The reader will also recognize that other constructive suggestions have already been brought up in Chapters 6 and 7 on composites, where new combinations of materials from the separate plastic "conti-

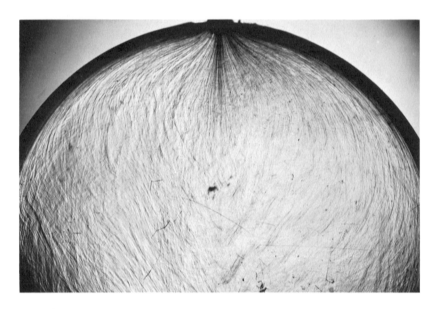

FIGURE 10-2.   Part of test disk (moldgate at top). *Injection* molded of two grades of poly(methyl methacrylate) of slightly different average refractice indices. (About 3×).

nents'' were encouraged. Other ideas will suggest themselves if the reader will review phases (Chapter 4) and surfaces (Chapter 5).

These few examples serve to illustrate the point that thoughtful interpretation of the results from a thorough resinographic investigation often can lead to new ideas and new designs that would never have emerged otherwise. This is the creative part of resinography. All of the detective work, the long search for facts and evidence, is analytical. When properly done, it leads to a solving of the mystery and to a sense of satisfaction in a successful closing of the case. But the usefulness of all these facts does not stop there. Once the *cause* of a structural defect or an optical aberration is uncovered, the effects can be projected deliberately in new directions, to arrive at new designs and new materials. Here a little reflection can be worth more than a thousand additional pictures, for it can produce something highly original and useful.

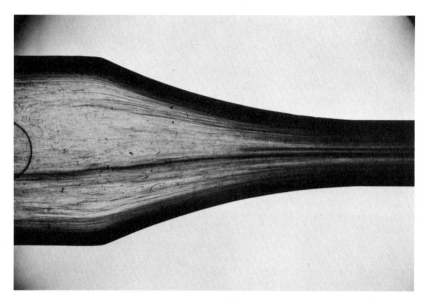

FIGURE 10-3.   Part of a tensile test bar. Composition and treatment as in
Figure 10-2. (About 3×).

SUMMARY

The purpose of this chapter is to show that information gained in analytical investigations need not remain in cold storage until the next similar problem comes along. Instead, it may be directed into constructive or creative channels if the resinographer is alert to other needs and trains himself to think creatively. Examples are drawn from the application of information gained in trouble-shooting investigations of compression molding and injection molding. Having learned the causes of inhomogeneities in the molded parts, it became possible to accentuate those inhomogeneities in deliberate directions, to produce new and useful decorative effects and stronger, better-appearing, molded objects. The reader is encouraged to direct all he learns from his own resinographic investigations into similarly original and constructive channels.

REFERENCES

1. Polymer properties, in *McGraw-Hill Encyclopedia of Science and Technology*, Vol. 10, p. 556, McGraw-Hill Book Co., New York, N.Y. 10020 (1971).
2. T. G. Rochow, Method for producing decorative articles of manufacture, *U.S. Patent 3,345,239*, Oct. 3, 1967, U.S. Patent Office, Washington, D.C., 20023
3. *Annual Book of ASTM Standards* and the annual *Index to ASTM Standards*, American Society for Testing and Materials, Philadelphia, Pa., 19103.

GLOSSARY

SYMBOLS FOR REFERENCES

ASTM   *Glossary of ASTM Definitions*, 2nd ed., American Society
           for Testing and Materials, Philadelphia, Pa. 19103 (1973).

EPS&T *Encyclopedia of Polymer Science and Technology*,
           Interscience Div., John Wiley and Sons, Inc.,
           New York, N.Y. 10016 (1966).

Mc-H   *McGraw-Hill Encyclopedia of Science and Technology*,
           McGraw-Hill Book Co., Inc., New York, N.Y.
           10036 (1960).

M-W   Merriam-Webster, *Webster's Seventh New
           Collegiate Dictionary*, G. and C. Merriam Co.,
           Springfield, Mass. 01101 (1969).

P-S   *Polymer Science*, 2 vols., A. D. Jenkins, ed.,
           North-Holland Publishing Co., Amsterdam (1972).

W   *Webster's Third International Dictionary*,
           unabridged, G. & C. Merriam Co.,
           Springfield, Mass. 01101 (1961).

+ = addition by authors
− = omission by authors
± = more or less

aberration   Any characteristic that results in image degradation. Such characteristics may be chromatic, spherical, or astigmatic, and can result from design or execution, or both.   *ASTM* ±

aberration, chromatic   An aberration in a lens or lens system resulting in different focal lengths for radiation of different wavelengths. The dispersive power of a single positive lens focuses light at the blue end of the spectrum at a shorter distance than light at the red end. An image produced by such a lens will show color fringes around the border of the image.   *ASTM* −

aberration, spherical   The zonal aberration of a lens referred to an axial point.   *ASTM* −

adduct   A chemical addition product.   *M-W*

adsorption   A physical process in which molecules of gas, of dissolved substances, or of liquids, adhere in an extremely thin layer to surfaces of solid bodies with which they are in contact.   *ASTM*

alloy   A metallic mixture of two or more elements to produce (1) a single phase (solid solution), (2) two or more solid solutions (phases) of varied composition and distribution, or (3) eutectic or eutectoid mixture (constant composition of two phases, uniformly distributed). Today the term alloy is also applied to resins, polymers, and plastics.

amorphous   Noncrystalline; devoid of a regular, cohesive structure.

analyzer   See polariscope.

anisotropic, optically   Having visible properties which vary with changing direction through the specimen.   *ASTM* ±

antiglare device   An accessory or method for reducing the intensity of reflection or the amount of non-image-forming light. Non-image-forming light may be reduced by coating air surfaces of lenses with a low-refractive film. In photography especially, glare from reflecting surfaces is removed by covering the lens with a polarizing film and turning it to the optimum position. See **glare**.

artifact   A spurious image which does not correspond to the true microstructure of the original specimen.   *ASTM* ±

artificial   Produced or effected by man to imitate nature; man-made.   *M-W* ±

asbestos   A group of fibrous minerals which occur as veins in the massive body of natural hydrous silicates of serpentine or amphibole and have specific heat-, fire-, and solvent-resistant properties according to species and variety.   *ASTM* ±

atactic   Having an unordered succession of monomer units of the two steric configurations. Such a polymer is not crystallizable *EPS&T* ±

attitude   An acquired predisposition to respond in a consistent way toward a given class of objects; a persistent state of readiness to react to a certain object or class of objects, not as they are, but as they are conceived to be.   *ASTM*

attribute   Any quality or characteristic descriptive of a stimulus. *ASTM*

behavior   Changes in properties of a substance with time, temperature, irradiation or illumination, humidity, or other environmental factors.

brand name   See **trademark**.

centrifugation, equilibrium   A method for determining the distribution of molecular weights by spinning a solution of the specimen at such a speed that the molecules of the specimen are not removed from the solvent but are held at a point where the (centrifugal) force tending to remove them is balanced by the dispersive forces caused by thermal agitation.

ceramography   The correlation of composition and treatment with properties and behavior, structure, and morphology of ceramics of all kinds.

chromatography   The separation, especially of closely related compounds, by allowing a solution or mixture to seep through an adsorbent (as clay, gel, or paper) so that each compound becomes adsorbed in a separate, often colored, layer.   *M-W* +

chromophore   A group of elements that gives rise to color in a molecule.   *M-W* +

chrysotile   The hydrous magnesium silicate mineral $Mg_3 Si_2 O_5$ (OH). Of all asbestos minerals, chrysotile is most commonly used in textiles, paper, board, and kindred materials. Because it contains water of crystallization which is lost at elevated temperatures, along with strength and some other properties, it is not as desirable as naturally anhydrous species of asbestos for use at such elevated temperatures.   *ASTM* $\pm$

cis   See **stereoisomer, cis.**

cleanliness   The quality or state of being free from dirt, contamination, admixture, encumbrance, incompletion, lack of skill (with respect to the specimen, substratum, superstratum, or optical system); orderliness.   *M-W* $\pm$

collimator   A device for controlling a beam of radiation so that its rays are as nearly parallel as possible.   *ASTM* $\pm$

colloid   A system of at least 2 phases, including a continuous liquid plus solid, liquid, or gaseous particles so small that they remain in dispersion for a practicable time.

composite   A material made up of distinct parts which contribute, either proportionately or synergistically, to the properties of the combination.

composition   The qualitative and quantititive statement of constituents. The source of information is analysis, and the derived facts need analytical interpretation.

configuration   The structural makeup of a chemical compound, especially with reference to the spatial relationship of the constituent atoms.   *M-W* $\pm$

conformation   The morphological disposition of a molecule in its environment, e.g., the coiling of a macromolecular chain in a poor solvent and the uncoiling in a good solvent.

contrast   Regarding perception, the ability to differentiate various components of the object's structure by different intensity levels in the image.   *ASTM* $\pm$

crystal   A solid composed of atoms, ions, or molecules arranged and cohered in a pattern which is periodic in three dimensions, except for occasional dislocations.   *ASTM* $+$

crystallinity, degree of   Ratio or percentage of crystallized phase(s) based on total material (or, perhaps, on crystallizable material only).

crystallizability   The inherent potential of a substance to arrange its units in a periodic, cohesive structure (crystal).

cues to depth (of object)   Hints or illusions of depth between near and distant points given by shadows, apparent size of similar objects, and perspective. Stereoscopy (effective use of both eyes) is very important in experiencing the third dimension. For relatively near objects the two eyes are effective; for distant objects, field glasses are needed; for close or microscopic objects, a binocular–binobjective microscope is required.

depth of field   The thickness of the object space that is simultaneously in acceptable focus.   *ASTM*

depth of focus   The thickness of the image space that is simultaneously in acceptable focus.

description   An image or impression in terms of morphology and structure. The description may be in words and numbers; photographs, patterns, and drawings; diagrams and spectra. Descriptive tools include macroscopy and microscopies; light, electron and x-ray scattering, diffraction or absorption; nuclear magnetic resonance, etc.   *ASTM*

differential thermal analysis (DTA)   A method for plotting (usually automatically, with two thermocouples bucking each other) the difference in temperature between a specimen and a neutral body against the temperature of the neutral body.   *ASTM ±*

e.g.   *Exempli gratia*, for example.   *M-W*

elastomer   A natural, synthetic, or artificial polymer which at room temperature can be stretched repeatedly to at least twice its original length and which after removal of the tensile load will immediately and forcibly return to approximately its original length.   *ASTM*

electron spin resonance (ESR)   Apparatus and method similar to NMR, but adapted to the much greater magnetic moment of an unpaired electron (trapped radical) compared with any nuclear moment.   *E*

emulsion   A stable dispersion of one liquid in another, generally by means of an emulsifying agent which has affinity for both the continuous and discontinuous phases. There is strong evidence that the emulsifying agent, discontinuous phase, and continuous phase together can produce another phase which serves as an enveloping (encapsulating) protective phase around the discontinuous phase.

epitaxy   The oriented growth of a crystalline substance on a substratum of the same or different crystalline substance.

equilibrium   A state of dynamic balance between the opposing actions, reactions, or velocities of a reversible process.   *ASTM*

experience   The conscious perception or apprehension of reality or of an external, bodily, or psychic event.   *M-W* ±

experimentation   Testing a hypothesis based on interpretations of observations or other experience with the sample.

field (of view)   The visible portion of the object.   *ASTM* ±

filler   A relatively inert material added to another (e.g., a plastic) to modify its strength, permanence, working properties, other qualities, or cost (as an extender).   *ASTM* ±

floc   A loose, open-structured mass produced in a suspension by the aggregation of minute particles.   *ASTM* ±

focus   A point at which rays originating from a point in the object converge or from which they diverge, or appear to diverge, under the influence of a lens or diffracting system.   *ASTM*

gauss   The unit of magnetic induction in the cgs electromagnetic system. The gauss is equal to 1 maxwell per square centimeter or 10^{-4} tesla.   *ASTM*

gel   A liquid containing a colloidal structural network that forms a continuous matrix and completely pervades the liquid phase. A gel deforms elastically upon application of shear forces less than the yield stress. At shear forces above the yield stress, the flow properties are principally determined by the gel matrix.   *ASTM*

generic term   Relating to, or characteristic of, a whole group or class.   *M-W*

glare   Harsh, uncomfortably bright light or, specifically, light which does not contribute to image formation.

glass transition The reversible change in an amorphous polymer or in amorphous regions of a partially crystalline polymer from (or to) a viscous or rubbery condition to (or from) a hard and relatively brittle one. *ASTM*

goniometry Measurement of the angle through which a specimen is rotated. *ASTM* ±

habit Characteristic mode of growth or occurrence of a crystal; characteristic assemblage of forms (free faces) at crystallization leading to a usual appearance. *M-W*

hardness The resistance of a material to deformation, particularly permanent deformation, indentation, or scratching. *Note:* Different methods of evaluating hardness give different ratings because they are measuring somewhat different characteristics and quantities of the material. There is no absolute scale for hardness, so that to express hardness quantitatively each type of test has its own scale of arbitrarily defined hardness. *ASTM*

hardness, impression The result of an imprint or dent in the specimen by the indenter of a hardness-measuring device. *ASTM*

hardness, Knoop That which is measured by calibrated machines to force a rhomb-shaped, pyramidal diamond indenter having specified edge angles under specified conditions, into the surface of the material under test and to measure the long diagonal after removal of the load. *Note:* The microhardness Knoop tester uses a relatively small load (1–1000 gf) to measure surface hardness. *ASTM* ±

hardness, Mohs' A scale of hardness first devised by Friedrich Mohs (1839), using pure minerals of selected species ranging from talc (softest) to diamond (hardest). The number of standards is now revised from 10 natural minerals to 15, including natural, artificial, and synthetic abrasives: fused silica (7), garnet (10), fused zirconia (11), fused alumina (12), silicon carbide (13), boron carbide (14). *M-W* ±

hardness, scratch A microhardness (of a surface) obtained by moving a diamond point (e.g., corner of a cube) across a surface of the specimen. *ASTM* ±

i.e. *Id est*, that is. *M-W*

illumination   The act of supplying light to the object by reflection, transmission, or at a grazing angle. In microscopy, each of the first two types has several kinds, axial bright-field, oblique or conical bright-field, dark-field, phase-amplitude. Conical bright-field illumination can be either "critical" or Köhler's illumination.

illumination, critical   The formation of an image of the light source in the object field.   *ASTM*

illumination, Köhler's (or Koehler's)   A method of microscopical illumination first described by A. Köhler, in which an image of the source is focused in the lower focal plane of the condenser.   *ASTM*

illumination, oblique   Illumination from light inclined at an oblique angle to the optical axis.   *ASTM*

imagination   The act or power of forming a mental image of something not present to the senses or never before wholly perceived in reality.   *M-W*

infrared (IR)   Pertaining to the region of the electromagnetic spectrum lying beyond the red, having wavelengths from 750 nm to a few mm.   *ASTM*

infrared (IR) spectroscopy or spectrometry   A method for observing or plotting the wavelengths in the electromagnetic spectrum lying beyond the red from about 750 nm to a few mm.   *ASTM* ±

in situ   In the natural or original position.   *M-W*

intangible separation   Distinction between individuals (such as different phases or separate particles) without physical movement (tangible separation); effective microscopical resolution.

interface   A surface forming a common boundary of two bodies, spaces, or phases.   *M-W*

intrinsic   Belonging to the essential nature or constitution of a thing.   *M-W*

intrinsic viscosity   The limiting value of reduced viscosity as concentration in a solvent approaches zero. The IUPAC Committee of Nomenclature has recommended the expression "limiting viscosity number" for intrinsic viscosity, and the concentration is generally expressed as grams per milliliter.   *ASTM* ±

isomer   A compound, radical, ion, or nuclide that contains the same number of atoms of the same elements but differs in structural arrangement and properties. See also **stereoisomer**.   *M-W*

isotactic   Pertaining to a type of polymeric molecular structure containing a sequence of regularly spaced asymmetric atoms arranged in like configuration in a polymer chain ("head-to-tail," etc.) Isotactic (and syndiotactic) polymers are crystallizable.   *EPS&T*

isotropic, optically   Having the same visible properties in all directions.   *ASTM*

lac   A resinous substance secreted by a scale insect and used chiefly in shellac.   *M-W*

lacquer   A coating formulation which is based on thermoplastic film-forming material dissolved in organic solvent, and which dries primarily by evaporation of the solvent. Typical lacquers include those based on lac, nitrocellulose, other cellulose derivatives, vinyl resins, acrylic resins, etc.   *ASTM*

lamella   A thin, flat scale or part.   *M-W*

lot   A collection of units of a product from which a sample is to be drawn and inspected to determine conformance with the criteria of acceptability, and is to be accepted or rejected as a whole.   *ASTM* ±

macromolecule   A large to giant molecule of a polymer.

macroscopy   The interpretive use of the naked eye, or with a magnification of no greater than 10 ×.

mass spectrometry (MS)   An instrument that is capable of separating ionized molecules of different mass/charge ratio and measuring the respective ion currents.   *ASTM*

material   Relating to, derived from, or consisting of matter.   *M-W*

memory   The power or process of reproducing or recalling what has been learned and retained especially through associative mechanisms; persistent modification of structure or morphology resulting from treatment of a material.   *M-W* ±

metal   A ductile, fusible, opaque, typically lustrous, crystalline substance which conducts electricity. The ductility (plasticity) is by slippage in the crystallographic directions of each crystal grain.   *M-W* ±

metallic whisker   A fiber composed of a single crystal of metal.   *ASTM*

metallography   That branch of science which relates to the constitution and structure and their relation to the properties, of metals and alloys.   *ASTM*

metallurgy   The science and technology of metals.   *M-W*

microdynamometer   An instrument for measuring mechanical force and observing the change in microscopical appearance of a small specimen.

microscope   An optical instrument capable of producing a magnified image of a small object.   *ASTM*

microscopic   Very small, pertaining to a very small object or its fine structure. A microscopic object requires microscopical examination to be adequately visible.   *ASTM*

microscopical   Pertaining to a microscope; pertaining to the use of a microscope.   *ASTM*

microscopy   The science of the interpretive use and applications of microscopes.   *ASTM*

molding   The pressing of powder or grains to form a compact unit.   *ASTM* +

molding, compression   The method of molding grains or powder in a confined cavity by applying pressure and usually heat. *Note*: This method is usually used to form objects that become thermoset.   *ASTM* +

molding, injection   A method of forming plastic objects from grains or powder by fusing the plastic in a chamber with heat and pressure and then forcing part of the mass into a cooler chamber where solidification takes place. *Note:* This method is commonly used to form objects from thermoplastics.   *ASTM* ±

molecular weight(s)   The sum of the atomic weights of all the atoms in a molecule. Atomic weights (and therefore molecular weights) are relative weights, arbitrarily referred to an assigned atomic weight for carbon. The molecular weight of a pure, unimolecular substance is as constant as the relative standard (carbon). Theoretically, at least, pure dimers, tetramers, and other pure oligomers have constant weights. *Mc-H*

molecular weight, number-average, $\overline{M}_n$ The result by any method determining molecular weight in which every molecule present produces the same effect, regardless of size. *Mc-H* ±

molecular weight, viscosity-average, $\overline{M}_v$ The result of measuring molecular weight by the method of intrinsic viscosity. The resulting value, $\overline{M}_v$, is generally between the weight average, $\overline{M}_w$, and the number average, $\overline{M}_n$.

molecular weight, weight-average, $\overline{M}_w$ The result by any method for determining molecular weight in which every molecule contributes in proportion to its size or weight, as in the measurement of light-scattering.

molecular weights, distribution The correlation between various macromolecular weights and their proportions in the total specimen. The resulting plot shows the extremes and average of molecular weight and the variation from the expected shape of the curve.

molecule The smallest particle of an element or compound capable of retaining identity with the substance in mass. *M-W* ±

monochromator A device for isolating radiation of one or nearly one wavelength from a beam of many wavelengths. *ASTM* ±

monomer A unimeric substance capable of reacting to form a polymer (or, at least, an oligomer). *ASTM* ±

montage An assemblage of separate pictures. *M-W* ±

morphology The shape and size of a line, an area or a volume; the texture or topography of a surface; the habit of a crystal; the distribution of phases in a system (material). Form may mean any of these or it may pertain to structure, so does not have a specific meaning unless given one. *ASTM*

nuclear magnetic resonance (NMR) Concerning radio frequency-induced transitions between magnetic energy levels of atomic nuclei. The instrument consists essentially of a magnet, radiofrequency accelerator, sample holder, sweep unit, and detector, capable of producing an oscilloscope image or line recording of an NMR spectrum. *ASTM* ±

object Something that is or is capable of being seen, touched, or otherwise sensed. *M-W*

oleoresins Nonaqueous secretions of resin acids dissolved in terpenic hydrocarbons that are produced or exuded from the intercellular resin ducts of living trees, especially the conifers, and accumulated in the wood of weathered limbs and stumps. *ASTM*

oligomer A polymer consisting of only a few monomer units such as a dimer, trimer, tetramer, etc., or their mixtures. *ASTM* ±

paracrystalline Having a highly defected, pseudocrystalline structure. *PS*

phase A visibly separate, but not necessarily separable portion of a system.

photography The production of images on a sensitized surface by the action of light or other radiant energy, especially that by standard processes, e.g., selection of photosensitive materials, exposure, development, fixation, magnification, labeling, etc. By photographical means, many of the other attributes of visibility may be changed. Photography is therefore listed as a separate attribute. *M-W* ±

photomicrograph An image produced on a sensitized surface by means of a microscope (light, electron, etc.) at a magnification over $10 \times$. *ASTM* ±

picturization An imaginary model or depiction of a visual conception of an idea or belief.

pigment The fine, solid particles used in the preparation of paint, ink, etc., and substantially insoluble in the vehicle. Asphaltic materials are not pigments except when they contain substances substantially insoluble in the vehicle in which they are used. *ASTM*

plaster A plastic system of solid particles which are free to move in a liquid. Old-fashioned plaster consisted of slaked lime, water, sand, and perhaps hair. Plaster of Paris consists of calcium sulfate hemihydrate, and water. The $CaSO_4 \cdot 1/2 \; H_2O$ sets to $CaSO_4 \cdot 2H_2O$ (gypsum).

plastic, *n.* A material which at some stage in its manufacture was shaped by flow. The term excludes metals, alloys, plasters, ceramics, waxes, and such plastic materials. *ASTM* ±

plastic, *adj.* Capable of being deformed continuously and permanently in any direction without rupture, under a stress exceeding the yield value. *ASTM*

plasticizer A substance incorporated into a material to increase its workability, flexibility, distensibility, etc. *ASTM* ±

plasticizer, external A substance incorporated into a polymer, resin, plastic, or fiber to increase its workability, flexibility, distensibility, etc. *ASTM* +

plasticizer, internal A comonomer or other reactant with the principal monomer, which increases the workability, flexibility, distensibility, etc., of the resultant polymer (i.e., the copolymer). *ASTM* ±

polar See **polars**.

polarimeter (1) An instrument for determining the amount of rotation of the direction of vibration of polarized light by the specimen; (2) an instrument for determining the amount of polarization of light by the specimen or in the illuminating beam. *M-W* ±

polariscope An instrument consisting essentially of a pair of polars: the **polarizer** for polarizing the beam illuminating the specimen and the **analyzer** for analyzing the effect, if any, of the specimen on the polarized light. At least one of the polars should be rotatable for obtaining crossed and uncrossed polars. Either the polars should be simultaneously rotatable or the specimen should be rotatable between polars.

polarized light Radiation in the visible range, which vibrates in only one direction, i.e., perpendicular to the direction or propagation.

polars Generally, a pair of polarizing elements. See **polariscope**. A polar may be a nicol (split calcite prism with parts cemented by medium of properly different refractive index) or a polarizing film such as Polaroid®, or some other polarizing device.   *ASTM* ±

polymer   A substance consisting of molecules characterized by the repetition of one or more types of monomeric units joined together by chemical bonds.   *ASTM* ±

polymorphism   The ability of a single substance to exist in more than one crystalline phase.   *ASTM* ±

polyphase   Having or producing two or more phases.   *M-W*

Portland cement   The product obtained by pulverizing clinker, consisting essentially of hydraulic calcium silicates.   *ASTM* −

preparation of the specimen   The process of making a part of the sample ready for observation or test, especially according to a standard procedure such as prescribed by the American Society for Testing and Materials.

profilometer   An instrument for drawing and measuring the ups and downs detected by a razor-sharp edge traveling over a surface such as that of a fiber.

properties   Characteristic qualities or quantities which serve to identify a material in a unique way, either as a peculiarity or as an attribute common to members of a class.   *M-W* ±

proximity   Closeness; distance between object and eye or objective of a microscope.

putty   A doughlike material consisting of pigment and vehicle, used for sealing glass in frames, and for filling imperfections in wood, metal surfaces, etc.   *ASTM* ±

radiation   (1) Emission or transfer of energy in the form of electromagnetic waves or particles; (2) the electromagnetic waves or particles *Note*: In general, nuclear radiations and radio waves are not considered in this vocabulary, only optical radiations (photons) of wavelengths lying between the region of transition to x-rays (1 nm) and the region of transition to radio waves (1 mm).   *ASTM*

radical   A group of atoms that is replaceable by a single atom which is capable of remaining unchanged during a series of reactions, or which may show a definite transitory existence in the course of a reaction.   *M-W*

replica   A reproduction of a surface of a material, e.g., a plastic. *ASTM* ±

replica, negative   A replica which is obtained by the direct contact of the replicating material with the specimen, so that the contour of the replica surface is reversed. *ASTM* −

replica, positive   A replica, the contours of which correspond directly to the surface being replicated, i.e., elevations on the surface are elevations on the replica. *ASTM*

resenes   The constituents of rosin which cannot be saponified with alcoholic alkali, but which contain carbon, hydrogen, and oxygen in the molecule. *ASTM* −

resin   A transparent or translucent, generally yellowish-to-brown solid or semisolid material which looks, feels, and conchoidally fractures like rosin, shellac, and amber. The term resin was later applied by Baekeland to describe his product of a phenol plus an aldehyde, whether alcohol-soluble, as shellac and rosin (non-polymeric), or practically insoluble in all physical solvents, as amber (polymeric). Today the term is also applied to other man-made polymers, whether insoluble (thermoset) or soluble (thermoplastic). *M-W* ±

resin, natural   A solid, organic substance, originating in the secretion of certain plants or insects, which is thermoplastic, flammable, nonconductive of electricity; breaks with a conchoidal fracture (when hard) and dissolves in certain specific organic solvents, but not water. *ASTM*

resinography   The science of morphology, structure, and related descriptive characteristics as correlated with composition or condition and with properties or behavior of resins, polymers, plastics, and their products. *Note:* Emphasis is on the *description* of such materials to provide a better understanding of *why* as well as *what* properties are obtained and of *how* the material is produced. *ASTM*

resolution   The fineness of detail revealed by an optical device. Actually, it is the separation of two points of the object on two different receptors, separated by at least one receptor on the retina of the eye. Resolution, including the acuity of the eye, is usually specified as the minimum distance by which two lines in the object must be separated before they can be revealed as separate lines in the image. It is customarily said that the smaller the distance, the greater the resolution (and the greater the required resolving power). *ASTM* ±

resolving power   The ability of a given lens system, including that of the eye, to reveal fine detail in an object. *ASTM*

rosin   A specific kind of natural, unimeric resin obtained as a vitreous, water-insoluble material from pine oleoresin by removal of the volatile oils or from tall oil by the removal of the fatty acid components. Rosin consists primarily of tricyclic, monocarboxylic acids having the general empirical formula $C_{20}H_{30}O_2$, with small quantities of compounds saponifiable with boiling alcoholic potassium or sodium hydroxide, and some unsaponifiable matter. The three general kinds of rosin are: (1) "gum" rosin from the oleoresin of living trees; (2) "wood" rosin from the oleoresin in dead wood such as stumps and knots; (3) tall oil rosin from tall oil. *ASTM* ±

rubber (natural)   The elastic substance obtained by coagulating the milky juice of any of various plants (as of the genera *Hevea* and *Ficus)* and prepared as sheets and then dried; called also caoutchouc, India rubber; chemically, essentially polyisoprene. *M-W* +

sample   The portion or unit(s) selected to represent the lot. *ASTM*

saturation, color   The attribute of color perception that expresses the degree of departure from the gray of the same lightness. All grays have zero saturation. *ASTM*

saturation in organic chemical compounds   Containing no double or triple bonds. *M-W* ±

schlieren   Regions of varying refraction in a transparent medium often caused by pressure or temperature differences and detectable especially by photographing the passage of a beam of light. *M-W*

shellac   Purified lac usually prepared in thin orange, yellow, or bleached white flakes or shells by heating and filtering; the lacquer prepared by dissolving shellac in alcohol. *M-W* +

size   Pertaining to textiles, paper, leather, plaster, and other porous surfaces—any of various glutinous materials as preparations of rosin, resin, glue, starch, etc., which fill the pores and thus reduce the absorption of ink, paint, or other liquid. *M-W* + *ASTM*

sliver   Bundles of noncontinuous or short-length fibers that have reached that stage of their fabrication into yarn wherein they are parallel and overlapping and have no twist.

specimen   An individual unit on which a test (examination) is made. The specimen may be representative or nonrepresentative of the sample and lot.   *ASTM* +

spectrometry   A method based on designation of the wavelengths within a particular portion of a range of radiations or absorptions, e.g., ultraviolet (UV) emission or absorption spectrometry. *ASTM* ±

spectrophotometer, recording   An instrument for measuring and automatically charting the intensity of radiation versus wavelength in a spectrum.   *M-W* +

spectroscopy   Concerning an instrument for dispersing radiation into a spectrum for visual observation of emission or absorption. *ASTM* ±

spherulite   A polycrystalline body of myriads of long (generally tiny) crystals radiating from a common center to form a sphere (or disk). Rotated between crossed polars, a spherulite shows a practically stationary Maltese cross, corresponding to those zones which always contain crystals in positions of extinction.   *M-W* + *ASTM*

staple   Natural fibers or cut lengths from filaments.

stereoisomer   An isomer in which atoms are linked in the same order but differ in their arrangement.

stereoisomer, cis   A stereoisomer in which atoms or groups of atoms are arranged on the same side of a chain of atoms.   *M-W*

stereoisomer, trans   A stereoisomer in which atoms or groups of atoms are arranged on opposite sides of a chain of atoms.   *M-W*

structure   The manner of construction from constituent parts such as atoms, radicals, or molecules in configuration and conformation in the liquid, glassy, rubbery, or crystalline state and in the phases of a system.   *ASTM*

substantive   Having a specific affinity for another substance, e.g., a fiber of cellulose or wool.

surface   The exterior or upper boundary of an object or body.   *M-W*

surface, specific   The surface per unit weight (or volume). *ASTM* ±

syndiotactic   Stereoregularity having alternately one and the other of the two enantiomorphic configurations ("head-to-tail–tail-to-head"). Syndiotactic, as well as isotactic polymers, are crystallizable. *EPS&T*

synergistic   The capability of discrete parts to act cooperatively, so that the total effect is greater than the sum of the effects taken independently.   *M-W* ±

synthetic   Relating to or involving a man-made product corresponding to a natural thing or the derivative.

tall oil   A generic name for a number of products obtained from the manufacture of wood pulp by the alkali (sulfate, Kraft) process.   *ASTM* −

thermogravimetric   Of or relating to measurement by weight of a product by heating.   *M-W* ±

thermoplastic, *n.*   A material which can be softened repeatedly by heat or solvents or extramolecular plasticizers, and hence can be etched by appropriate solvents.   *ASTM* ±

thermoset, *n.*   A material which was hardened by chemical reaction and cannot be re-softened by heat. It is relatively insoluble and can be etched only by appropriate chemical reagents.   *ASTM* ±

topography   The description of a surface including its relief and the position of its features.   *M-W*

tow   Man-made fibers—a twistless, multifilament strand suitable for conversion into staple fibers or sliver, or for direct spinning into yarn.   *ASTM*

trademark   A device (as a brand name) pointing distinctly to the origin of merchandise to which it is applied and legally reserved to the exclusive use of the owner as maker or seller. Trademarks as words are capitalized and are followed by a mark such as ® (registered), followed by the descriptive name, e.g., Polaroid® camera, photosensitive material or polarizing film; Fiberglas® fibrous glass.   *M-W* ±

trade name   The name or style under which a concern does business, e.g., DUPONT for E. I. duPont de Nemours and Company. The trade name by itself or with a device such as a surrounding oval may also be registered by the concern with the government.   *M-W* ±

trans   See **stereisomer, trans**.

tristimulus system of colorimetry   The method of matching a color by measuring the amounts of the three stimuli, X, Y, and Z.   *ASTM* ±

ultraviolet (UV)   Pertaining to the invisible region of the electromagnetic spectrum, usually referring to the region from 200–380 nm, but including 10–380 nm.   *ASTM* −

unimeric   Pertaining to a single molecule which is not monomeric, oligomeric, or polymeric, e.g., saturated hydrocarbons.

varnish   A liquid formulation that is converted to a transparent or translucent, solid film after application as a thin layer. Typical is oil varnish which contains resin and drying oil as the chief film-forming ingredients and is converted to a solid film primarily by chemical reaction.   *ASTM* ±

wax   Any substance physically resembling beeswax, a dull, crystalline solid, plastic when warm; composed of a mixture of esters, cerotic acid, and hydrocarbons. Other waxes are from various animal, vegetable, mineral, or synthetic sources. Waxes differ from fats in being less greasy, harder, more polishable by rubbing, and in containing esters of higher fatty acids and higher alcohols, free higher acids or higher alcohols, and/or saturated hydrocarbons (e.g., ozokerite; paraffin).

whisker, metallic   See **metallic whisker**.

working distance   The depth of space between the surface of the specimen and the front surface of the objective lens (of the unaided eye).   *ASTM* ±

Ziegler-Natta catalysts   Initially consisting of an alkylaluminum compound together with a compound of the titanium group of the periodic table, a typical combination being triethylaluminum, and either titanium tetrachloride or titanium trichloride. Subsequently, an enormous variety of such mixtures is used in polymerization to provide stereospecificity (isotactic or syndiotactic). *EPS&T* ±

AUTHOR INDEX

SUBJECT INDEX

Numbers in italics refer to the Glossary, which begins on page 165.